THE RENDEL
CONNECTION
A Dynasty of Engineers

James Meadows Rendel

THE RENDEL CONNECTION

A Dynasty of Engineers

Michael R. Lane

Foreword by Rt. Hon. Margaret Thatcher F.R.S., M.P.

Quiller Press
LONDON

First published 1989
by Quiller Press Limited
46 Lillie Road
London SW6 1TN

ISBN 1 870948 01 7

Designed by Hugh Tempest-Radford Book Producers
Typeset by Halcyon Type & Design, Ipswich
Printed in Great Britain by Purnell Book Production Limited

Contents

Foreword
by
Rt. Hon. Margaret Thatcher, F.R.S, M.P.

Rt. Hon. Margaret Thatcher, F.R.S, M.P.

10 DOWNING STREET

LONDON SW1A 2AA

THE PRIME MINISTER

A 150th Anniversary is, by any standard, a great achievement and I send my congratulations to Rendel Palmer & Tritton.

They have had a distinguished history as one of the leading firms of consulting engineers involved in providing major harbours, roads, bridges and power stations all over the world.

These achievements include the design of the Howrah Bridge across the Hooghly River, in India, and, nearer home, the Chelsea Suspension Bridge, Waterloo Bridge and the Thames Barrier. I was also pleased to hear that last year the firm won a contract against tough international competition to design the Jamuna Bridge in Bangladesh.

Many things have changed over the last 150 years and Rendel Palmer & Tritton have witnessed many of them. But some have remained the same. For example in the 1830s James Meadows Rendel drew up documents for toll roads near Plymouth. Today they are still doing the same thing in several parts of the world.

Rendel Palmer & Tritton have much to be proud of over the last 150 years. I wish them continuing success into the 21st Century.

Margaret Thatcher

JULY 1988

Acknowledgements

In preparing the following History I am greatly indebted to the directors and members of staff both past and present, of Rendel Palmer & Tritton and other High-Point Group companies, for their helpful assistance in the writing of this book. This is especially the case with Chapters 8 – 12 which cover recent projects and events and which address the changing face of engineering consultancy. My gratitude is due especially to Peter Cox, the last Senior Partner of this great firm before incorporation and a past President of the Institution of Civil Engineers. His unique knowledge of events in the past two decades made possible the final chapters dealing with the Thames Flood Barrier and events leading to the formation of High-Point Rendel. Francis Irwin-Childs, Peter Cox's predecessor, provided a constant source of encouragement and much valuable information concerning the numerous major projects undertaken by the firm in the 1960s and 1970s, including the great docks at Seaforth, Liverpool and Portbury, Bristol, and the many nuclear and fossil fuel power stations designed for the C. E. G. B. and other clients.

Kempton Bedell-Harper, the Marketing Services Manager, assisted by Mr. J. Broadis, the firm's Librarian unearthed a great deal of valuable information for me from the company's archives, always dealing patiently with my many queries and providing great support and encouragement. I am indebted also to Miss D. Bayley, Librarian, Institution of Civil Engineers for access to various published works and to Mrs Eileen Jarvis, the former Editor of *Rendels News,* the firm's excellent House Magazine.

My very special thanks go to Miss Rosemary Rendel, Sir Alexander Rendel's grand-daughter and Sir George Rendel's daughter, and to Mr Keith S. Perkins and Mr A. W. Neal. Miss Rendel provided valuable information about the Rendel family and its many fascinating connections. Mr Perkins has an enormous wealth of knowledge concerning James Meadows Rendel's early life and especially his 'Floating Bridges'. Mr Neal made freely available his draft manuscripts dealing with the life and works of James Meadows Rendel and Sir Frederick Palmer and I have drawn extensively on both gentlemen's work.

I am grateful to Dr R. F. D. Porter-Goff of Cambridge University Engineering Department and to Mr Ronald H. Clark, the well known Norfolk author, for reading my manuscript and making many helpful corrections and improvements.

Finally, my thanks go to Juliet Edwards who patiently and expertly put my manuscript on a word processor, to Hugh Tempest-Radford who designed and produced the book, to Jeremy Greenwood, my publisher, and to my wife, Margaret, all of whom have shown great kindness and understanding during the preparation of this book.

Michael R. Lane
November, 1988

1

James Meadows Rendle
The Formative Years 1799-1822

Few travellers in the mid-18th century disagreed with Horace Walpole's verdict on the state of our roads. A popular writer and younger son of Sir Robert Walpole, later becoming the 4th Earl of Orford, he continually complained about their execrable condition. The deterioration had been progressive since the Romans left in AD 407, the relatively small amount of improvement and repair work undertaken thoughout the Middle Ages being financed by big landowners and the monasteries rather than by the state. In 1555 Parliament intervened passing legislation that put the problem on a national basis, although another 250 years elapsed before practical solutions were found following the introduction of new methods of road construction by the innovators Thomas Telford (1757-1834) and John Louden McAdam (1756-1836).

The Statute for Mending Highways of 1555 made the inhabitants of each parish responsible for the repair and upkeep of the roads within the parish. Every person holding land having an annual value of £50 and every person owning draught horses or a plough had to provide, at appointed times, 'one wain or cart with oxen, horses or other cattle, and all other necessaries meet to carry things convenient for that purpose, and also to have two able men with the same'. Every householder in the parish and every labourer other than servants hired on a yearly basis, had to work on the road, or alternatively arrange a substitute in his stead. The statute required that the work was supervised by a surveyor and carried out during a period of four days each of eight hours duration. In 1562 the period was increased to six days, and fines were imposed on all those who refused to perform the duty. The Act further required that 'the constables and churchwardens of every parish shall yearly, upon the Tuesday or Wednesday in Easter week, call together the parishioners and shall then choose and elect two honest persons of the parish to be surveyors and orderers for one year'. The surveyor's main responsibilities were to survey all the roads and bridges under his jurisdiction at least three times a year and report their condition to the Justices; ensure all ditches were clear and that no trees were overhanging or interfering with the roads; see that the statutory duty was performed at the right time in a proper manner; and maintain accounts for submission to the Justices.

The system did not work well especially in those small parishes having long stretches of important main roads in the area. In such cases local resources were frequently inadequate to keep the highways in good repair. Farmers in particular were often reluctant to provide statute labour, considering the Act a legal burden from which no benefit could be derived, and it soon became apparent that more work could be done with three hired teams than with five statute teams'. General Highways Acts passed in the 18th century enabled people liable to statute duty to pay money for the upkeep of roads rather than providing the labour. Increasingly, the maintenance of bridges and other structures raised all sorts of problems and many disputes arose as to who was responsible for the work.

Eventually Parliament authorised the setting-up of Turnpike Trusts, giving the trustees the power to raise capital for road improvements on the security of the charges levied on road users as they passed through the toll gates. The trustees were charged with the appointment of paid officials to carry out the day-to-day work of the trust, comprising a clerk, treasurer and surveyors. The surveyors were employed either on a full-time basis, or on a part-time contract, thereby enabling them to engage in other forms of gainful employment. Salaries ranged from a few shillings a week to £500 per annum. The larger trusts provided their surveyors with a horse, and in the case of the Bristol trust where the much sought after John McAdam was appointed surveyor-general in 1816, a horse and carriage was provided.

James Rendle (1767-1838), living in the little Devonshire village of Drewsteignton, nestling in the beautiful valley of the River Teign, was a typical part-time West Country road surveyor responsible for the important main roads radiating from the market town of Okehampton in the directions of Exeter and Newton Abbot. Rendle was a common name in the West Country, although various other forms of spelling were frequently used – Rendel and Rendell in Devonshire and Rundle in Cornwall. It was by no means unusual for different generations of the same family to vary the spelling, as we shall see as our story unfolds. Early documents indicate that James Rendle's family had been yeomen farmers in the Okehampton district for several generations, and James combined his work as a road surveyor with that of an estate agent, valuer and auctioneer with an office in nearby South Tawton. He was described as a man of more than ordinary stature, having a massive frame and an intelligent face, plain in manners and reserved in conversation, but conveying the impression of excellent judgement and a well regulated mind.

In 1794 he married Jane Downie, the young widow of Thomas Downie a local farmer, by whom she had a son, Samuel. Jane was the eldest girl of John and Jane Meadows' five children, and was generally regarded as coming from a socially superior background to that of the Rendle family. Her father, John (1732-1791) was a distinguished architect, for many years associated with James Wyatt, the designer and improver of many fine Georgian houses. Although Meadows, a Member of the Society of Arts and a Liveryman, gave his address as Tufton Street in Westminster, he spent a great deal of his time in the West Country. He was responsible for re-building parts of the original Arlington Court, near Barnstable, the home of the Chichester family for over five centuries. He supervised the construction of the house at Hurstbourne Park for the 2nd Earl of Portsmouth, and in 1779 carried out extensive rebuilding work at Hartland Abbey in North Devon. He became a trusted friend and adviser to John Parker, the 2nd Baron Boringdon, later created the 1st Earl of Morley, who lived at Saltram House, Plympton, the largest mansion in Devonshire. Fanny Burney, a well-known 18th century novelist, described the house as, 'one of the most magnificent in the Kingdom, its

Drewsteignton Church, Devonshire. An early lithotint from nature by W. Sprait.

view is noble.' The interior decoration afforded one of the best examples of Robert Adam's work, and the house contained an outstanding collection of paintings by Sir Joshua Reynolds, who was born nearby and had an intimate association with the house. Meadows was a close friend of Dr. Samuel Simmons, physician to King George III and Queen Charlotte. When the royal couple visited Saltram House in 1788, both Simmons and Meadows were amongst the 100 distinguished guests accommodated in the house. The connection between Lord Morley, a great improver and local benefactor, and John Meadows was to have profound consequences for the latter's as yet unborn grandson.

James Meadows Rendle was born in 1799.

James Rendle and his wife continued to live in her former matrimonial home at Thornbury farm, near Drewsteignton and in due course they produced a son and five daughters. The son, named James Meadows Rendle, was born in 1799. Little is known about his early childhood other than that his mother, described as a woman of considerable attainments, undertook his early education. Subsequent family papers indicate that from an early age the boy showed a keen interest in his father's work and learnt from him the rudiments of civil engineering. When 12 years old James was sent to live with an uncle in Teignmouth to be instructed in the work of the millwright, whilst continuing his education at a local county school. Several members of the Rendle family lived in the district at this time, and it is not now clear which uncle took young James in hand. Uncle John was a builder, valuer and Port Reeve at Teignmouth which then enjoyed a brisk trade with Nova Scotia and Newfoundland. Timber was imported and locally manufactured goods, mainly from the Holbeam edge-tool mill at Ogwell were shipped out. The water-wheel driven mill was owned by another member of the family, and Holbeam fish hooks and fisherman's knives were highly regarded and in great demand in the Newfoundland fishing grounds. Another uncle, William Rendle, was a ship owner and ship builder, who amongst other things built the quay at Teignmouth on land leased from Lord Clifford.

Saltram House — The Earl of Morley's home — 'one of the most magnificent in the kingdom.'

Between 1810 and 1813 during James Meadows Rendle's childhood years spent at Drewsteignton and Teignmouth, Thomas Telford, the undisputed doyen of the civil engineering profession, undertook several assignments in the West Country advising Turnpike Trusts and others on the construction of new roads, bridges and other structures. Although McAdam's name was perhaps better known in the area in relation to roads, Telford was acknowledged as the master of bridge design and construction. Prior to this time the majority of his road works had been in the Scottish Highlands, for he had not yet commenced his famous London to Holyhead road. In the early years of the 19th century it became acknowledged that bridges were a key element in the construction of new roads, and Telford's advice on the subject was in great demand throughout the Kingdom. He was consulted by Lord Morley on various matters and the two men became firm friends, Telford having stayed at Saltram House on several occasions. It is not surprising therefore that we learn that James Rendle senior had also met the great man in the course of his duties as a road surveyor.

Opinions differed as to the best methods of constructing and repairing roads. McAdam who came into prominence in 1798 as the Falmouth district surveyor, emphasised the need to use materials producing a firm foundation suitable for the heaviest vehicles coupled with proper drainage. In his much publicised book, *The Present System of Road Making,* he advocated, 'a road ought to be considered as an artificial flooring forming a strong, smooth and solid surface at once capable of carrying great weight and over which carriages may pass without meeting any impediment'. He advocated a soft foundation in preference to a hard one, claiming that a road placed upon a hard substance, such as rock, wore more quickly than when placed upon a soft substratum. He recommended that a road should be as flat as possible, consistent with rain water being able to run off the surface easily. Generally he built his roads 18 ft wide, three inches higher in the centre than at the sides. He insisted upon the proper training of his surveyors and that they should always use stones of equal size. They were at all times required to carry a pair of scales and a six ounce weight to check each delivery of stones. He employed large numbers of men and women stone breaking, using hammers with 15 ins long handles and a 1 in. face. The stones were laid evenly on the road to a depth of 6 ins, followed a few weeks later by a further 6 ins thick layer. Perhaps McAdam's greatest contribution to the nation was political, successfully campaigning that the public roads became the responsibility of Government, financed out of taxes for the benefit of everyone.

Telford, like McAdam, advocated keeping the roads as level as possible, maintaining as a general rule that the gradient should not exceed 1 in 30. Unlike McAdam, he recommended a hard, solid foundation using broken rock carefully placed as a covering on a solid stone foundation with the central working part of the road as hard as possible. He used stone blocks 7 ins by 4 ins for his foundations with stone chips wedged between them. These were covered by a layer of stones 6 ins deep, topped by a further layer of small stones or gravel. The working parts of his roads were also about 18 ft wide, with the addition of lesser surfaces each 6 ft wide on either side. Locally available materials largely determined the type of road constructed, although some Trusts went to the expense of importing better quality materials from other areas.

James Meadows Rendle succeeded in obtaining employment as a student surveyor on Telford's staff.

By 1815 it was the ambition of every aspiring engineer to gain experience under Telford's guidance. Having decided to follow in his father's footsteps, young James Meadows Rendle, with his father's help, succeeded in obtaining employment as a student surveyor on Telford's staff. Initially, he was engaged surveying roads in North Devon, and after this probationary period, during which his enthusiasm and conscientiousness were noted, he was sent to Telford's London headquarters to learn the art of draughtsmanship. Then followed his first major assignment. Together with William Provis he was given responsibility for the preparation of detailed drawings for a suspension bridge across the Mersey at Runcorn.

In 1818, Telford had completed his plans for a suspension bridge across the Menai Straits 100 ft above high water with a central span of 550 ft connecting the Isle of Anglesey with the Caernarvonshire mainland. During the preceding 25 years several prominent engineers, including John Rennie, Telford's principal competitor, had been trying to solve the problem of improving communications between London and Dublin via the port of Holyhead. It was the hazards of crossing the Menai by ferry which travellers most dreaded. Despite parliamentary support for a bridge, it was the Admiralty who ruled against the proposed schemes. They insisted that navigation through the Straits must not be obstructed and that vessels of the largest size must be able to pass freely with masts erect. They would not even countenance any obstruction of the channel by the centering which would be required to build an arch bridge. Initially these restrictions had seemed crippling, for they necessitated an arch of such unprecedented height and span, precluding accepted methods of construction. The first stone of the bridge's main pier at Ynys-y-Moch was laid in August 1819 by Rendle's former colleague William Provis, who had been appointed Telford's resident engineer. Concurrently with the Menai operation Telford built the smaller Conway Suspension Bridge and both were opened to the public in 1826.

During the period the Menai Bridge was in abeyance, Telford had been approached about a scheme to bridge the Mersey at Runcorn. Once again he decided that a suspension bridge was the only solution. The American engineer, James Finley, was the first to build successfully a suspension bridge across a major river at Jacob's Creek in Pennsylvania, but with the exception of a few footbridges, the principle was at that time untried in Britain. Together with Bryan Donkin, Peter Barlow and others, Telford carried out an exhaustive series of experiments into the tensile strength of malleable iron using a special Bramah hydraulic press at William Brunton's Chain Cable Works for the purpose. As a result of these experiments he designed a suspension bridge of spectacular proportions having a centre span 1000 ft long and 70 ft above high water, with two side spans each of 500 ft. His original design proposed suspending the bridge platform by means of four sets of Brunton's laminated iron cables. In the light of subsequent knowledge engineers have been forced to the conclusion that it was just as well the bridge was never built. It has been calculated that the design would have offered too little resistance to lateral wind pressure acting on the 1000 ft span.

While Telford was conducting his Runcorn experiments, unbeknown to him a naval officer, Captain Samuel Brown R.N. (1780-1852), was working on similar lines, which he eventually used in the construction of the Union Bridge across the River Tweed near Berwick. Brown abandoned the use of laminated cables in favour of a chain composed of long flat iron links and pins. As a young officer Brown had seen distinguished service at sea in the war against the French, and in 1809 made his name as an engineer by demonstrating the advantages of replacing hemp cordage with iron chain cables for ships' anchors. The Admiralty quickly took up the idea and ordered four warships be fitted with chain cables. The idea was also soon adopted by the Merchant Fleet. As soon as Telford heard of Brown's work on bridges, he contacted him with the result that the two men pooled their knowledge and built a model of the proposed Runcorn Bridge. Although their experiments were a great success, and despite the unanimous acceptance of the design by the bridge committee, the project came to nothing. Undoubtedly, the abandonment of the Runcorn Bridge was a great disappointment to young Rendle, but the experience gained soon proved of inestimable value to him. His associaton with Captain Brown continued for twenty years, but events often brought the two men into conflict.

The abandonment of the Runcorn Bridge was a great disappointment to young Rendle.

Shortly after the Runcorn affair in January 1819, a meeting took place at the King's Head Tavern in Cheapside. Henry Maudslay, the distinguished engineer and millwright, took the chair, supported by his colleague Joshua Field. A resolution was passed, 'that a society be formed consisting of persons studying the profession of civil engineering'. Thereafter, the initiative was taken by one of Telford's most promising employees, Henry Robinson Palmer, a former Bryan Donkin apprentice. He was assisted by James Ashwell, one of Donkin's chief assistants. In 1820, the society was renamed the Institution of Civil Engineers and Thomas Telford was elected its first President, thereby launching it as the premier body in the realms of engineering. Without Telford's great influence the new body would almost certainly have foundered, but he was determined it should succeed as a pool of knowledge and experience to promote the advancement of the profession. In 1828, the Institution was granted a Royal Charter signifying the national recognition of civil engineering as a profession of honour and repute.

Reverting to James Meadows Rendle's career, we learn that his next assignment took him to Scotland, where Telford was preparing plans for the landing piers for a proposed ferry across the River Tay between Dundee and Newport. On this occasion Telford was in competition with the eminent Scottish engineer, Robert Stephenson (1772-1850), the celebrated lighthouse builder. Eventually, Telford's scheme was adopted and he was awarded a contract to build the piers and supervise the supply

Telford's Tay Ferry Pier — circa 1820.

of twin-hulled steam powered ferry boats having a large centrally mounted paddle wheel. The *Union* was launched in James Brown's Yard in Perth in 1821 and the *George IV* followed two years later. Both vessels were developments of William Symington's (1762-1831), *Charlotte Dundas* which was first demonstrated on the Forth and Clyde Canal in 1803. Symington had in fact built his first steam boat as early as 1788, but the *Charlotte Dundas* had the distinction of being the first vessel to employ a piston rod coupled to a crankshaft by means of a connecting rod to drive the paddle wheel.

Whilst engaged working on the Tay Ferry, Rendle met John Mitchell (1775-1824) and his son Joseph (1803-1883). John had started life as a stonemason in Forres and eventually became Divisional Superintendent under Telford in his capacity as Engineer to the Highlands Roads and Bridges Commission. With Telford's assistance his son received a good education at the Inverness Academy, later becoming one of his articled pupils, and upon his father's premature death, Joseph took over the Highlands Roads and Bridges. Such was Telford's fame in Scotland at this time, his fellow countrymen honoured him with a Fellowship of the Royal Society in Edinburgh. Thereafter he encouraged his young assistants to take every opportunity of attending the Society's meetings, and for a time Rendle and Mitchell became great friends visiting Edinburgh together. Subsequent events sadly soured the relationship and some years later, at Inverness, the two men were in open conflict.

Mitchell served the Highlands Roads and Bridges Commission for 18 years during which time he built 40 churches as the Church of Scotland representative. Then he became Engineer to the Scottish Fisheries Board, responsible for the design and maintenance of harbours all round the Scottish coast. In the 1840's his influence undoubtedly assisted Rendle in obtaining important work at Leith docks. In 1844, he surveyed and laid out the Scottish Central Railway via Bathgate. Thereafter, he devoted the remainder of his working life to railway projects in the North of Scotland.

He is perhaps best remembered to-day as the the first member of the Institution of Civil Engineers to establish minutes of the Proceedings of the Institution.

It seems probable that another young man Rendle met whilst working in Scotland was James Nasmyth (1808-1890), the inventor of the steam hammer. The son of an eminent landscape painter, Nasmyth had demonstrated a talent for foundry work and metal working since the age of twelve. In later years he wrote in words of great wisdom, which are equally true to-day.

> I had the good luck to have as a school companion the son of an ironfounder. Every spare hour that I could command was devoted to visits to his father's ironfoundry, where I delighted to watch the various processes of moulding, iron melting, casting, forging, pattern making, and other smith's and metal work. I was only about 12 years old at the time, but I used to lend a hand, in which hearty zeal did a good deal to make up for want of strength. I look back to the Saturday afternoons spent in the workshop of that small foundry as an important part of my education. I did not trust to reading about such things, I saw and handled them, and all the ideas in connection with them became permanent in my mind. I also obtained there, what was of much value to me in later life, a considerable acquaintance with the character and nature of workmen.

At the age of 15, Nasmyth built his first steam engine and soon found a ready market for both working engines and sectioned demonstration models. He did some work for his father's friend, Patrick Miller, the patentee of the paddle wheel and at the age of 17 he built a steam carriage which was demonstrated in Edinburgh carrying eight people. This innovation was ahead of its time and sadly came to nought. In the same year he approached the Directors of the Forth and Clyde Canal Company with a scheme which would allow the movement of steam powered tugs and barges along the canal without damage to the banks. Nasmyth proposed that a chain cable be laid along the bed of the canal in such a manner that the boat's steam engine could warp its way along the chain. Initially, the chain was carried over rollers in the bow, passing over a sprocket wheel on the engine crankshaft and returned to the water at the stern. Later, Nasmyth proposed the use of a twin-hulled boat with the steam engine and boiler mounted centrally between the two hulls. This substantially shortened the amount of chain out of the water at any time, greatly improving the steering and buoyancy. Although Nasmyth's proposals were 'courteously declined' the idea was revived in the 1860's and extensively used on canals and waterways on the Continent and in the United States. It is an interesting speculation whether Rendle discussed chain haulage with Nasmyth before he left Telford's employment, for he must have been aware that the high winds and strong currents encountered on the Tay were already proving too much for the *Union* and the *George IV*. Perhaps the idea of the 'Floating Bridge' ferry was already beginning to crystallise in Rendle's mind ten years before he was able to put it into practice.

No record of Rendle's attitude of mind or of his aspirations and ambitions whilst working in Scotland has survived. We are left to speculate that perhaps he was homesick and longed to return to his family and the softer climate of the West Country. Perhaps he saw little future for himself working in an environment dominated by Telford's Scottish Presbyterian associates, with whom he was not always on the best of terms. There were signs also that Telford was getting old and tired and tending more and more to leave regional affairs to his senior colleagues resident in the area. No single assistant had emerged as the natural successor and it appeared that Telford was resigned to the fragmentation

In 1822 James Meadows Rendle decided to return to Devonshire, putting up his plate as 'James Meadows Rendel, Civil Engineer'.

of his remarkable business once he had gone.

In 1822, at the age of 23, James Meadows Rendle made one of the most important decisions in his life and decided to return to Devonshire, putting up his plate at 7, Boon's Place in Plymouth as 'JAMES MEADOWS RENDEL, CIVIL ENGINEER'. The reasons for adopting the 'RENDEL' spelling of his name are not now known, but one can be sure his mother's strong influence played some part in the decision. It is clear that this new venture had the full blessing and approbation of his former employer, for until the controversial Clifton Bridge contests in 1829, Rendel enjoyed Telford's goodwill and recommendation.

The reasons for adopting the 'Rendel' spelling of his name are not clear.

2
James Meadows Rendel, Civil Engineer, Plymouth 1822-1838

JMR's first commission was to carry out a series of road surveys.

James Meadows Rendel's first commission as an independent civil engineer was to carry out a series of road surveys on behalf of Turnpike Trusts in North Devon. His father's local connections supported by letters of recommendation from Telford undoubtedly helped him initially. He was a personable young man, having an open honest face, articulate, and over 6 ft tall. In spite of his age he had little difficulty communicating with potential clients and he quickly gained the confidence of employers and their workers alike. The respect he quickly earned ensured that henceforth he was kept fully employed.

The history of the long hours and extreme hardship endured by engineers and surveyors at this time, often travelling great distances over rough ground in all kinds of weather, is well documented. It is said that JMR, as he became generally known to his friends and associates, often refused to ride horseback when out and about doing field work, claiming that he was too heavy for the average horse. Undoubtedly, he had a strong constitution and was robust, competent and totally dedicated to his work.

Plymouth, then administratively separate from the adjacent naval base at Devonport, was a good centre in which to establish his business. The town was rapidly growing and trebled its population in the first half of the 19th century. Maritime trade especially was on the increase and the building of Sir John Rennie's breakwater made Plymouth Sound one of the safest harbours of refuge in the country. The need locally for improved roads, bridges, port and dock facilities, fresh water sewers, and all things of concern to the civil engineer had never been greater.

JMR also found the social and cultural life of the district to his liking. He gave several erudite papers before the Plymouth Institute, which later became known as the Plymouth Athenaeum. This establishment had similarities with the famous Birmingham Luna Society, and the Wedgewood and Darwin families, leading members of the Luna Society, later became connected by marriage with the Rendel family. Socially, JMR associated with the town's professional set. His closest friends were Samuel Elliot, a surveyor, later to become his brother-in-law, George Wightwick and John Foulston, leading architects in the district, and the Greaves family, who had earlier sold their gracious family home to Lord Nelson's mistress, Lady Emma Hamilton. George Wightwick was bestman at JMR's wedding in 1828 and he named his second son, George Wightwick Rendel, after him.

George Wightwick, distinguished architect and life long friend of James Meadows Rendel after whom he named his third son.

The idea of a suspension bridge across the Tamar at Saltash was under consideration by a group of noblemen and landed gentry living in the area before JMR decided to return to the West Country. Indeed awareness of the existence of this ambitious scheme so soon after his acute disappointment at Runcorn may have been a deciding factor in the decision to leave Telford and set-up on his own account. Once in Plymouth he lost no time in preparing plans for a spectacular suspension bridge in the full knowledge that he was in competition with Captain Samuel Brown. On August 4th 1822, having submitted his plans, he attended a public meeting at Callington on the Cornish side of the Tamar. He spoke out convincingly with the result that a number of resolutions favourable to him were passed. Although he had won the active support of his grandfather's influential friend, the Earl of Morley, he had to wait patiently for eighteen months for a decision. On March 2nd 1824, the Committee of Management of the Duchy of Cornwall, under the Chairmanship of Lord Morley, formally announced that the scheme had been put in abeyance due to difficulties with the Treasury.

Lord Morley had been actively involved in several development schemes in both Devon and Cornwall for many years. In 1807, he had been awarded the Gold Medal of the Society of Arts. Founded in 1754, its full and proper title was the Society, later to become The Royal Society for the Encouragement of Arts, Manufacturers and Commerce, and the medal was awarded to

> 'the person who shall produce to the Society an account, verified by actual experiment, of his having gained the greatest quantity of land from the sea, not less than 50 acres, on the coast of Great Britain and Ireland and that the work was commenced subsequently to 1st January 1800.'

Morley had engaged the Birmingham born engineer, James Green, to supervise the reclamation of 75 acres at Chelson Bay on the east bank of the estuary of the River Plym near where it flowed into the Cattewater at Laira. This necessitated the construction of a 970 yds long embankment standing 18 ft above the level of the highest recorded spring tide and having a base width of 91 ft. Water penetration was prevented by a covering of limestone paving on the sea side of the embankment, which stood on a water-tight clay puddle. Water coming from the surrounding higher ground and from springs in the reclaimed land was collected in a drain and carried off through a 3 ins thick oak plank sluice. A similar embankment was constructed by the Earl at Charleton near Kingsbridge, enabling a further 40 acres of land to be recovered for the benefit of his tenants.

Prior to 1807, access to the Earl's residence at Saltram House, from the Plymouth side, had involved a circuitous route crossing the River Plym at Longbridge on the main Plymouth to Exeter road. His Lordship obtained an Act of Parliament enabling him to build what was described as a 'Flying Bridge' across the river at Laira on the Cattewater connecting Plymouth with Pomphlett Point, south of the Chelson Meadows reclamations. Not only was this intended to improve greatly access to Saltram House, but also to encourage the development of a more direct route to Modbury and Kingsbridge. The 'Flying Bridge' was probably originally developed by the military and comprised a simple raft arrangement tethered by a long hawser to an anchor in the river bed and allowed to swing diagonally across the water with the current. The raft was then hauled back to its starting point by a manually operated capstan on the shore, allowing the operation to be repeated as wind and tide permitted. Lord Morley's ferry, large enough to carry 3 or 4 carriages, was operated somewhat differently, having the capstan on the ferry and an iron chain replacing the rope

The Earl of Morley's 'Flying Bridge' built in 1807 and replaced by JMR's Laira Bridge in 1827.

hawser. Delays were frequent and it proved unpopular with travellers. None of the local coaches used the service and it was condemned by many as a hindrance to navigation. Lord Morley himself became increasingly frustrated by the delays caused by wind and tide, and when he learned of the opening of Captain Brown's suspension bridge across the River Tweed near Berwick, began to explore the possibility of a similar bridge being built to replace the unsatisfactory 'Flying Bridge'.

Lord Morley had been much impressed by JMR's plans for the Tamar Bridge.

Clearly, His Lordship had been much impressed by JMR's plans for the Tamar Bridge and by his performance at the Callington meeting. After discussing the matter with Telford, he summoned Rendel to Saltram House on October 2nd 1822, suggesting that he adapt his Tamar proposals to provide a suspension bridge across the Cattewater at Laira, between Great Prince Rock and Pomphlett Point. JMR accepted the challenge with great enthusiasm, and in less than two months, on November 22nd, he submitted his revised plans. The Earl of Morley then requested a committee of eight 'civil and practical' engineers to comment upon the proposals. The committee included such eminent names as Marc Brunel, Henry Maudsley, Joseph Bramah, Bryan Donkin and James Walker, the architect and engineer responsible for London's Vauxhall Bridge. At this time scientists and mathematicians were making great strides in assessing the natural laws underlying the functions of arches, beams, cantilevers and stiffened suspension bridges, and there existed no more distinguished body of men able to endorse Telford's young proteges plans. Byran Donkin was the mathematician in the group, and he had a great influence upon JMR's subsequent work.

Quoting JMR's own words in his paper given before the Plymouth Institution in 1829:

> The drawings being approved, an Act of Parliament was obtained in the Session of 1823 for carrying the plan into effect. Subsequently, however,

> circumstances occurred to occasion the abandonment of the site first proposed, being unfavourable to the erection of a bridge on the principle of suspensions and the original intention was relinquished. In the Session of 1824, another Act was therefore obtained, repealing the Act of 1823, so far as related to the suspension bridge, and extending its powers to meet the erection of the present structure.

Events leading to this change of plan were complicated, involving not only the technical considerations referred to by JMR, but also local vested interests and clashes of personality. A Welsh engineer, Roger Hopkins, the builder of the Pen-y-Darren Tramway, made a spirited attempt to promote the idea of a wooden bridge. He gained the support of a number of local people, including Daniel Asher Alexander, Surveyor to the Prince of Wales and Trinity House, but the scheme was eventually abandoned due to the difficulty and cost of deep driving wooden piles. The Admiralty, who usually opposed all bridges over tidal waters because of their threat to navigation, for once raised no objections. The principal opponents formed a committee led by Sir William Elford, the Recorder of Plymouth and the owner of a wharf on the Laira, and a Doctor Joseph Cookworthy, who later played a vital rôle in saving JMR's life following a fall from the bridge. Other objectors were the Trustees of the Plymouth and Exeter toll road who saw their traffic diminishing if the bridge was built, and the Dartmoor Railway Company, whose proposed line crossed the approach road to the bridge on the Plymouth side. Lord Morley had difficulties persuading the Modbury Road Trustees to implement a separate Act he had obtained for road improvements on the eastern side of the bridge and in spite of his wishes and influence as Chairman of the Trust, the trustees snubbed young Rendel by appointing John McAdam as its engineer, claiming that JMR was too expensive. His Lordship also had to face local criticism that the bridge, when built, would be the property of one man rather than a corporate body.

JMR had other problems of a personal nature to overcome before Lord Morley finally gave the go-ahead for the construction of the bridge. In the early stages, when a suspension bridge was still advocated, Captain Brown, who was then building the Brighton Chain Pier, wrote to Lord Morley accusing JMR of copying his plans. The young man was quick to reply, and in a letter to His Lordship he wrote:

> I conceive it will be quite necessary for me to remark that Captain Brown wishes to grasp at everything in the shape of suspension bridges and therefore of course feels sore at the idea of one being built without him. It appears to me that Captain Brown in the heat of the moment, forgot himself. How does he reconcile those parts of his letter where he first accuses me of making an exact transcript of his plans, or mode of forming the chains and then to assert that were I to get 30 ft of chain made agreeably to my drawings, he would prove to Your Lordship that the bridge would not stand twenty-four hours? Certainly, my Lord, this is a flat condemnation of his own patent, or clears me of what he intends to prove.

The opposition tried to make capital out of JMR's youthful age.

No doubt JMR felt completely vindicated and greatly encouraged when the group of eminent engineers reported that in their opinion Brown was wrong.

The opposition tried to make capital out of JMR's youthful age and limited experience. Even John Yolland of Merrifield, Lord Morley's agent and an adjudicator in the Society of Arts reclamation contest, wrote to His Lordship expressing his doubts. He referred to JMR as a 'theoretical engineer, never having had any experience in fixing coffer-dams or masonry of such a description or other works of this nature'. Undoubtedly, without Telford's constant behind-the-scenes support, coupled with Lord Morley's influence, Rendel would have had the

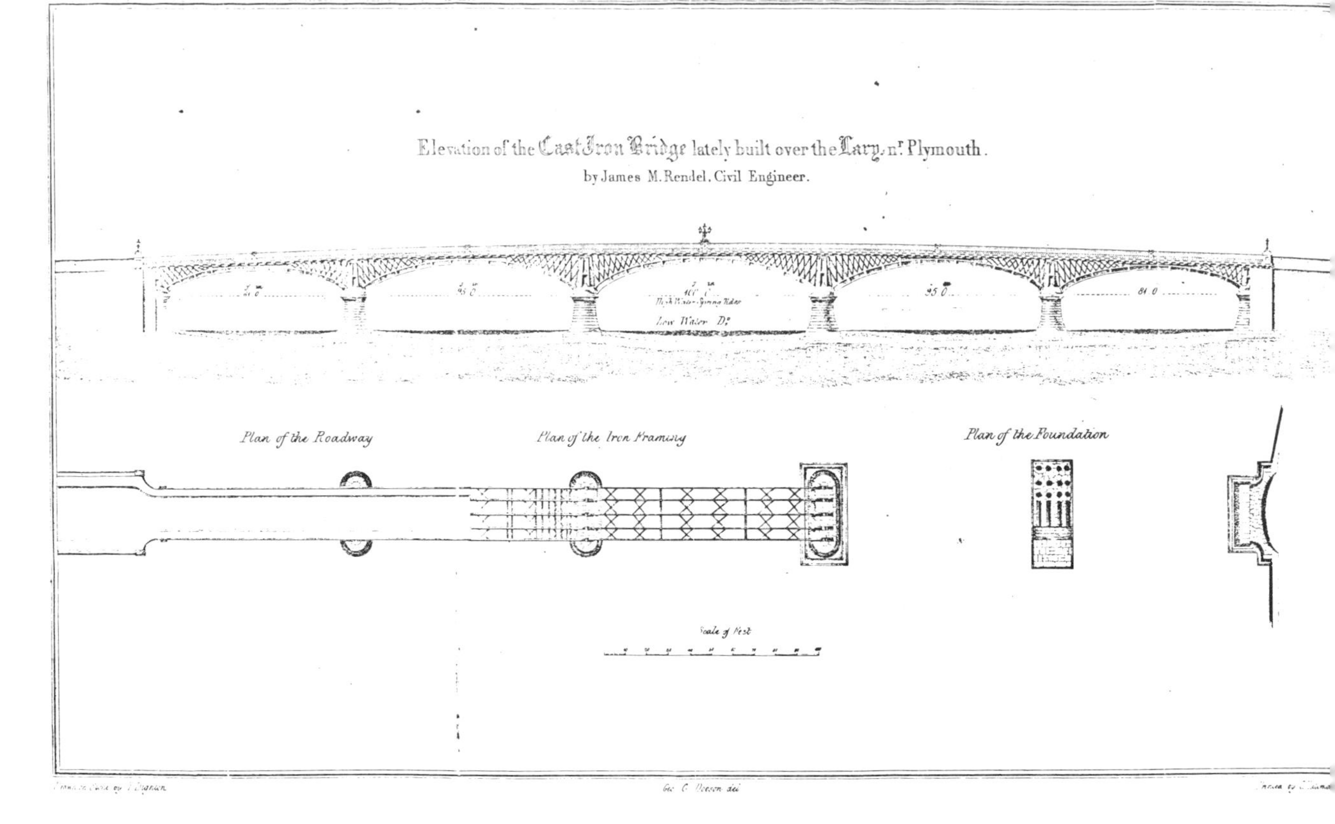

Elevation of JMR's Cast Iron Bridge at Laira, Plymouth, as illustrated in the proceedings of the Institution of Civil Engineers. Signed by George C. Dobson, JMR's brother-in-law.

greatest difficulty securing this prestigious job. Members of the 'Committee of Eight' also continued to give their active support and confirmed their recognition of his competence and professionalism by proposing him for full membership of the newly formed Institution of Civil Engineers in 1824.

Rendel's final design was for a bridge of five spans with cast iron arches.

Rendel's final design was for a bridge of five spans with limestone masonry piers and abutments, granite pier caps and cast iron arches and superstructure, and work was commenced on August 4th 1824. The centre arch was 100 ft long, the two intermediate arches each 95 ft long, and the two end arches each 81 ft long. The total length of the bridge within the abutments was 500 ft, and each arch was 22 ft above High Water Spring Tide Level. The total width of the carriageway and footways was 24 ft. John Johnson & Company and their associates, the Plymouth Granite Company, were in the initial stages awarded the contract for the foundations and masonry work. William Hazledine, (1763-1840) the Coleham and Plas Kynaston ironfounder, was awarded the contract for the cast iron arches and superstructure. He was the younger brother of John Hazledine, the Bridgnorth ironfounder, who built Richard Trevithick's first locomotive engines. William had known Telford since 1789 when both men were members of the Salopian Masonic Lodge and had been closely associated with him in business since the building of the Pont Cysyllte and Chirk aqueducts at the end of the 18th century. He was responsible also for the elegant ironwork of Telford's Waterloo Bridge at Betwys-y-Coed in North Wales, which still carries the busy A5 road over the River Conway. It was unthinkable that once JMR was committed to the use of an iron superstructure for his Laira Bridge anybody other than Hazledine should supply the ironwork.

JMR gave his justification for the final design of the bridge in great detail in his 1829 paper before the Plymouth Institution. The strait across which the bridge was to be built was only 550 ft wide and formed by limestone cliffs. Trial borings had revealed that the substratum of slate rock was some 80 ft below high water. It was overlain by softer deposits from river and tidal action to a depth of some 60 ft. Spring tides caused a surface velocity of the flow through this comparatively narrow channel of 3.5 ft per second and a rise of about 18 ft to 19 ft.

The need to have the bridge arches as large as economically feasible in order to minimise the effects of obstructing the flow which would have increased its velocity and deepened the channel was clearly of paramount importance. The cost of building adequate foundations for the piers and abutments that would have been necessary to carry large ponderous stone arches was excessive. JMR therefore decided to use cast iron because of its durability and strength combined with lightness. He quoted the recently completed iron bridges at Sunderland, Southwark and Tewkesbury as precedents.

JMR also appreciated that as the work proceeded it would be necessary to provide some form of bed protection in order to prevent scour adjacent to the bridge foundations. Experiments showed that the strong red clay found in the neighbourhood of the bridge was not eroded when exposed to a current of water moving at a velocity of 7 ft per second acting upon its surface. As soon as the bed of the river had scoured away to a depth at which he felt it expedient to commence bed strengthening, he deposited the red clay and calcareous spar mixed with small stones to a depth of 2 ft over the whole surface of the bed or the river. This deposit was then covered with rough blocks of limestone and spar to within 1 ft of the original bed of the river. The blocks once firmly embedded in the clay protected the lower softer strata from erosive action, thereby satisfactorily securing the bridge foundations.

In November 1824, a disastrous storm damaged the bridge works and breached the embankment at Chelson Meadow. Lord Morley's 'Flying Bridge' was torn from its mooring and flung ashore upside-down. JMR repaired the embankment with earth brought by barges, but almost immediately another gale breached it a second time. JMR was obliged to build a temporary tramway to transport earth and rubble in order to speedily repair the damage and the work was completed by the end of the month.

By March 1825 fifty men were at work on the bridge piling, but the contractors raised all manner of difficulties. This further delayed progress and eventually resulted in their withdrawal from the contract, leaving JMR with the sole responsibility for the work. During this difficult period Lord Morley showed signs of impatience with his engineer and at short notice he despatched JMR to the Midlands by stagecoach for consultations with various people. In Gloucester he had discussions with Telford, who was then engaged supervising the Gloucester & Berkley Canal on behalf of the Exchequer Loan Commission, and with a Mr McIntosh, one of Rennie's assistants. He met William Hazledine in Birmingham and returned with him to Shrewsbury, where a fixed price of £17 14s 0d per ton was agreed for the ironwork. Not only was Lord Morley concerned about rising costs, but he had promised the Modbury Turnpike Trustees that the bridge would be completed by 1826 and that compensation would be paid if completion were delayed.

On March 22nd 1825 the first caisson of a unique design developed by JMR was ready for launching.

The first 6 ton stone block was placed at the Pomphlett abutment on March 22nd 1825, and the first caisson of a unique design developed by JMR was ready for launching. It comprised a watertight box constructed from beech wood containing the masonry for the base of the piers. It was floated out over the piles, sunk in position and the caisson sides were then dismantled for reuse.

JMR also designed the diving-bell used for finishing off the pile heads to take the first courses of the masonry work for the piers and abutments. Much depended on the regularity with which the pile heads were levelled

JMR's first major work. The Laira Bridge, Plymouth, completed in 1827.

and great care was bestowed upon this work. Equally, the paving between the pile heads was of great importance in order to make a firm foundation for the piers. Instead of the usual metal structure, the diving-bell was made from two layers of elm board made water-proof with flannel soaked in beeswax and resin. It was reinforced with wrought iron and lined with lead. It measured 5 ft 6 ins in length, 4 ft 6 ins in width and 5 ft in height and was provided with two moveable seats and foot boards for the divers. On June 25th it was tested for the first time, but the air pump gave trouble and it was not put to work until July 4th. JMR himself went down the first time in order to give his men confidence and is on record as 'being very pleased with the results.'

Throughout the remainder of 1825 and into 1826 the work seems to have continued without further difficulty. On December 20th Sir William Elford's Plymouth Bank went into liquidation ending all opposition to the project. The piers and approach roads were completed by October 1826, but none of the ironwork had arrived on site. Hazledine wrote to Lord Morley explaining that there had been insufficient water in the Severn during the summer to enable barges carrying the ironwork to get down to Bristol. He assured His Lordship that he now had men at Stourport to transfer the ironwork from narrow boats to river barges and others at Bristol to transfer it to seagoing vessels for the passage round Land's End. By the end of October 100 tons had been loaded on the *Agramonia* at Bristol, and Hazledine's manager, Mr Stuttle, had arrived in Plymouth to supervise the erection. By the end of November 140 tons had arrived and a further 150 tons was in transit. Although Hazledine had failed to meet his delivery dates, JMR recommended to Lord Morley that the penalty of £100 per month should not be charged.

An accident occurred which nearly proved fatal to JMR.

During the erection of the ironwork an accident occured which nearly proved fatal to JMR. He fell off the scaffolding into the water and being burdened by a heavy coat and jack boots was submerged for some time before being rescued. Injudicious treatment by his rescuers almost extinguished the last glimmer of life, but fortunately the bridge's old adversary, Dr Joseph Cookworthy appeared on the scene in the nick-of-time. He managed to revive JMR and a close and cordial friendship developed between the two men. In later life JMR recorded that, 'whilst under the water, I experienced the most vivid retrospect of every circumstance of my previous life with a minuteness and fidelity marvellous.'

In February 1827 another tragedy befell the project. Stuttle caught a severe cold, and although at first he did not appear to be in danger, his condition worsened and JMR sent for his son, William. On February 23rd, before he could arrive, Stuttle died. William then took over his father's responsibilities and proved himself an energetic and competent manager. Inspite of the inevitable day-to-day problems, JMR felt confident that the bridge would be finished and ready for the formal opening by the end of June 1827.

With completion so near the most important problem became the inscription to commemorate Lord Morley's public spirited action. The Rt. Hon. George Canning, then Foreign Secretary, John Walter, the proprietor of *The Times*, the Earl of Carlisle and other scholars were consulted. Rendel wrote to Lord Morley requesting that his name might also be placed unobtrusively on the opposite abutment. His wish was fulfilled, for not only did his name appear on the bridge, it was also incorporated into the main inscription which read:

HUNC PONTEM
SENATUS AUCTORIATE SUSCEPTUM
NOVAS ET COMMODAS
VIAS RECLUDEN TEM
JOHANNES COMES DE MORLEY
SUIS SUMPTIBUS
STRUENDUM CURAVIT
OPUS INCHOATUM A.D. 1824
ABSOLUTUM A.D. 1827
J. M. RENDEL ARCHITECTO

On July 14th 1827 the bridge was formally opened amid great pomp and circumstance by Her Highness the Duchess of Clarence, later to become Queen Adelaide, the wife of the Lord High Admiral. She arrived from Exeter by carriage and together with the Earl and Countess of Morley drove across the bridge into Plymouth to be greeted by a huge crowd of spectators. Two weeks later all the workmen employed during the construction were given a splendid celebration dinner.

The final cost of the bridge was £27,000, sustantially in excess of JMR's original estimates, but the excellence of the design and execution of the work has never been questioned. With the exception of Sir John Rennie's Southwark Bridge over the Thames, it was the largest iron structure then in existence, and the Institution of Civil Engineers promptly awarded Rendel a coveted Telford Medal for his work. It survived exactly as built for over 130 years until replaced by a modern concrete structure in the 1960's. The southern abutment and a part of the original roadway remains virtually intact, and a low wall has been added containing the now sadly vandalised original inscription. Thus, whilst still under 30 years of age, James Meadows Rendel had established a reputation as being amongst the leaders of contemporary British bridge designers.

The Institution of Civil Engineers awarded Rendel a Telford Medal for his work.

A large proportion of the stone and slate used in the construction of the bridge and its approach roads came from the Cann Quarries, situated 5 miles from Plymouth on the north eastern side of the River Plym. Owned by Lord Morley and managed by his agent, John Yolland, His Lordship had ambitious plans for developing both the quarries and the means of getting their produce to Plymouth for shipment. Large deposits of China Clay, the decomposed felspar of granite, second only to the Cornish deposits in quantity and quality, had also been found in the area. China Clay was fast becoming the largest of Britain's raw material exports as it became increasingly in demand for paper manufacture, pharmaceutical products and cosmetics.

Morley realised he was sitting on great untapped wealth, and in 1826 he instructed JMR to prepare plans for a canal from the quarries to the river Plym at Marsh Hills. As soon as this scheme had been resolved, His Lordship turned his attention to the building of a railway from Plymouth to connect with the canal. Because of his heavy commitments at the Laira Bridge, JMR employed his friend Samuel Elliott to carry out the surveys and prepare the drawings, although he personally always signed them. In 1829, a 4 ft 6 ins gauge railway, financed entirely by Lord Morley, was opened from Rising Sun at Crabtree on the outskirts of Plymouth, to meet the canal at Marsh Mills. Later, in 1833, the canal was closed and in conjunction with the Plymouth and Dartmoor Railway, the line was extended along the canal towpath to the Cawn slate quarry. The line remained busy until 1855 and was eventually closed in 1900.

The idea of a railway was not new. Another politician, Sir Thomas Tyrwhitt, who had unsuccessfully spent huge sums of money trying to develop Dartmoor, obtained three Acts of Parliament authorising the Plymouth & Dartmoor Railway having its terminus at Princetown. In September 1823 the 23 miles of 4 ft 6 ins gauge single track were opened from Sutton Pool in Plymouth to King Tor, 1300 ft above sea level and only 14 miles from Plymouth as the crow flies. Financial difficulties overtook the company almost immediately and the Johnsons, originally awarded the contract for the foundations and masonry work at the Laira Bridge, effectively took control. Lord Morley had opposed Tyrwhitt's 1821 Bill, but as payment for the withdrawal of his opposition, the company agreed to build the Cann Quarry branch. Their precarious financial position was worsened by the collapse of the Plymouth bank and the bankruptcy of several directors leaving the responsibility for the implementation of the agreement in the hands of Lord Morley's solicitor, John Pridham. Although His Lordship instituted proceedings against Pridham and the company, these were dropped and he financed the line himself, the work being based on JMR's surveys. In 1847, the Marsh Mills to Cann Quarry branch was purchased by the South Devon Railway.

In the closing years of JMR's life he was retained by Lord Morley to survey and supervise the building of the adjacent Lee Moor Tramway connecting the China Clay mines with Plymouth. This ill fated line was closed only one month after its opening in 1854 following an accident on the Torracombe incline and was not re-opened until 1858. These episodes are typical of many in the history of the 4 ft 6 ins gauge railways of Devon and Cornwall. Today, production of China Clay at Lee Moor exceeds 500,000 tons annually and is piped to the Marsh Mills treatment works before shipment.

Other works with which JMR was concerned in 1827 were the completion of the new Toll Gate houses at the approaches to the Laira Bridge; the building of the new Cann Quarry wharves at Pomphlett Point; the design and building of a new inn at Pomphlett, to be used as a staging post for the Exeter to Plymouth stagecoaches; and the layout and construction of Plymouth race course at Chelson Meadow. He charged Lord Morley one guinea for planning Lady Morley's 'bathing place'. He regularly attended meetings of the various Turnpike Trusts with which Lord Morley was involved; he attended the newly formed Torpoint Ferry

Par Harbour, Cornwall. In mid-1
century JMR's first important har
constructed on an open beach.

Committee meetings; and he negotiated terms with Mr Church of Exeter, a stagecoach operator, who agreed to operate a regular service crossing the new Laira Bridge. At the Earl's request he waited on a Mr Knight of Poole in Dorset, who had expressed interest in acquiring the old Cattewater 'Flying Bridge'. Obviously, he found time also to indulge in some courting, for on January 31st 1828 he married Catherine Jane Harris, but that story belongs to a later chapter.

Later in 1828, JMR was approached by Sir Christopher Hawkins Bart, a Cornish landowner and owner of the small fishing harbour at Pentewan in Mevagissey Bay. He wished to gain control of the shipment of China clay mined in the St Austell district. As far back as 1821 a tentative survey of a route between the clay pits and Newquay had been carried out and work on the completion of Newquay harbour had been put in hand. JMR's report dated April 1829 recommended that a 15 acre reservoir be constructed at Pentewan to supply backwater through a canal or conduit in order to clear the basin and preserve the channel entrance. He proposed additionally a new fish pier and the construction of two new breakwaters of stone from the adjoining cliffs. This outer harbour would both provide shelter and prevent silting up. JMR claimed that if these works were carried out, they would make Pentewan one of the safest artificial harbours on the south coast. He further advocated an investigation into the possibility of a railway extension from the clay pits to the port and improvements to the roads in the district. Sir Christopher took no action inspite of severe gales having already threatened the viability of Newquay as a port. In 1843, after JMR had removed his practice to London and was fully committted with schemes of great national importance, Captain William Moorsom (1804-1863), the former engineer to the Birmingham

JMR claimed these works would make Pentewan one of the safest artificial harbours on the south coast.

& Gloucester Railway, was commissioned to re-examine Rendel's proposals. Moorsom's scheme was also aborted and the Cornish clay industries shipments were henceforth concentrated on the Falmouth area and later at Par, near St Austell.

The Austen family originating in Plymouth and the Treffry family of Fowey had for long been prominent in mining and various other trades in Cornwall. They shipped out tin, copper, clay, wool and fish, and Austen was also a successful brewer. The two families were united by marriage in the 18th century and their various business interests became even more closely intertwined. In 1838, Joseph Thomas Austen (1772-1850), then head of the family and High Sheriff of Cornwall, received the Royal Assent to change his name to Treffry and henceforth to bear the Treffry arms. Austen had extensive mining interests in the St Austell district, and all the coal he required for smelting had to be shipped from South Wales to the Port at Fowey. It had then to be transported 10 miles or so by mule train to the mines. Similarly, his finished products had to be laboriously carried back to the port for shipment. One day while standing on high ground overlooking St Austell Bay pondering the problem, he suddenly had the vision of developing nearby Par Sands as a port. Situated near the village of St Blazey, Par was an ideal location and he immedately contacted JMR whilst he was still engaged at Pentewan.

The idea of constructing a port on an open beach had been considered previously by Smeaton and other eminent engineers and condemned as impracticable. Austen met Rendel at Par on April 6th 1829, and together they proceded to take soundings along the bay. After initial doubts, Rendel agreed that a new pier could be built within the shelter of Spit Rocks. Between April 29th and May 1st he was busy from early morning until dusk taking soundings and making trial borings along the southern edge of the rocks. He proposed a 500 ft long pier containing an outer floating basin linked to the inner basin by a second canal within the pier. The 45 acre tidal reservoir was to be protected by substantial oak gates, and JMR claimed that vessels would be able to enter the harbour under canvas on almost any wind.

Concurrently, Austen was busy constructing a tramway to the mines incorporating a series of inclined planes on the steep hillside behind St Blazey, enabling him to bring the ore down to sea level by gravity. The empty wagons were then hauled back to the mine workings by stationary steam engines. He used the bed of a small adjacent stream to build a ¾ mile long canal to Par Sands, where he erected a new copper smelter. Upon their completion, much to Austen's delight, it was found that two horses towing the canal barges which JMR had designed could do the work of 300 mules and horses previously employed on the rough roads between the mines and Fowey.

Unfortunately, Austen proved to be extremely autocratic and difficult to please, and the two men soon quarrelled. Eventually JMR withdrew from the project and was compelled to take legal procedings in order to recover his fees amounting to £136. Despite severe problems with constantly swirling and shifting sand, Austen persevered, overcoming the problem by constructing a 1200 ft long breakwater. Basically, JMR's original plans were implemented, thereby providing quayside and mooring space for over fifty small merchant ships.

The Runcorn Bridge project had been a great disappointment to young JMR, the Tay Ferry essentially a failure, the Laira Bridge, although a great personal triumph when completed, had been achieved in the face of considerable opposition and frustration, and at Par he had come face to face with an extremely difficult client. Throughout these formative years he conducted himself with dignity and professionalism, strengthening his character and developing his business acumen, whilst earning the respect of his peers. Always a good communicator, self-assured and speaking his mind plainly, he never suffered fools gladly.

* * *

In 1753 a Bristol wine merchant named William Vick had bequeathed a sum of £1,000 to the Society of Merchant Venturers, the eminently respectable and extremely wealthy group of local traders who had for long boasted that they alone kept their trade independent of London. Vick's wish was that the money should be accumulated at compound interest until it reached £10,000 which he optimistically supposed would be sufficient to build a bridge across the River Avon at Clifton. By 1829 the advent of railways had already meant that Liverpool had supplanted Bristol as England's second city and the worthy citizens of Bristol decided to take steps to protect their trade and economic position. The legacy then totalled £8,000 and a committee was formed to determine what could be done to fulfil Vick's ambition. Within weeks land was purchased, designs for a new bridge were invited and it was agreed that any additional finance required would be raised by public subscription.

At this time young Isambard Kingdom Brunel (1806-1859) was staying in Clifton, recuperating from the accident in the Rotherhithe Tunnel which so nearly ended his career. He lost no time consulting Joseph Field and his father, who had earlier designed a suspension bridge for the French Government on the Ile de Bourbon. He paid a two day visit to North Wales and minutely examined Telford's Menai and Conway Bridges. On the closing date of the competition he submitted four designs of suspension bridge situated at different points along the Avon gorge. He took infinite care preparing his designs and his drawings were magnificent examples of draughtsmanship. The proposed spans varied between 870 ft and 916 ft, all greater than any suspension bridge yet built.

The Bridge Committee invited Telford to judge the merits of the designs submitted and to assist them in the final selection of a site. Somewhat surprisingly he summarily rejected Brunel's designs on the grounds that the spans were too great, stating that he considered 600 ft was the maximum safe span for a suspension bridge. Following the Great Man's rejection of all the submissions, the committee felt that they had no alternative but to ask Telford to submit his own design. Since described as a 'Gothic Folly', Telford proposed a bridge having a central span of only 360 ft with side spans each 180 ft long. The two massive supporting towers rose 240 ft from the floor of the gorge and were decorated in florid Gothic. Indeed, in retrospect, it is difficult to understand how such a design could have come from the mind which, only a few years earlier, had produced the graceful spans of the Menai and Conway Bridges.

PARTICULARS

DESCRIPTIVE OF

THE ACCOMPANYING DESIGN FOR

AN INTENDED

BRIDGE OVER THE AVON,

AT CLIFTON,

BY

JAMES M. RENDEL,

MEMBER OF THE INSTITUTION OF CIVIL ENGINEERS,

&c. &c.

PLYMOUTH:
ROWE, 9, WHIMPLE-STREET.

1830.

Title page of JMR's entry for the Clifton Bridge Contest, October 1830.

Brunel was furious and led a spirited opposition which succeeded in persuading the Bridge Committee to hold a second competition in October 1830. A total of twelve designs were submitted and one of these was presented by James Rendel. He proposed the exact site for his bridge and submitted very detailed plans for both the bridge and the approach roads. His highly professional accompanying report dealt at length with the strength of the suspension chains and towers and considered ways of preventing roadway undulation. He based his maximum loadings on a column of marching soldiers. He detailed no less than four mathematical investigations into the strength of chains involving Poissons Theory and recommended the burning oil treatment for the corrosion resistant protection of the chains. With respect to the strength of the masonry towers he recorded his own earlier experiments on the compression strength of stone. His estimated total cost was £92,885.

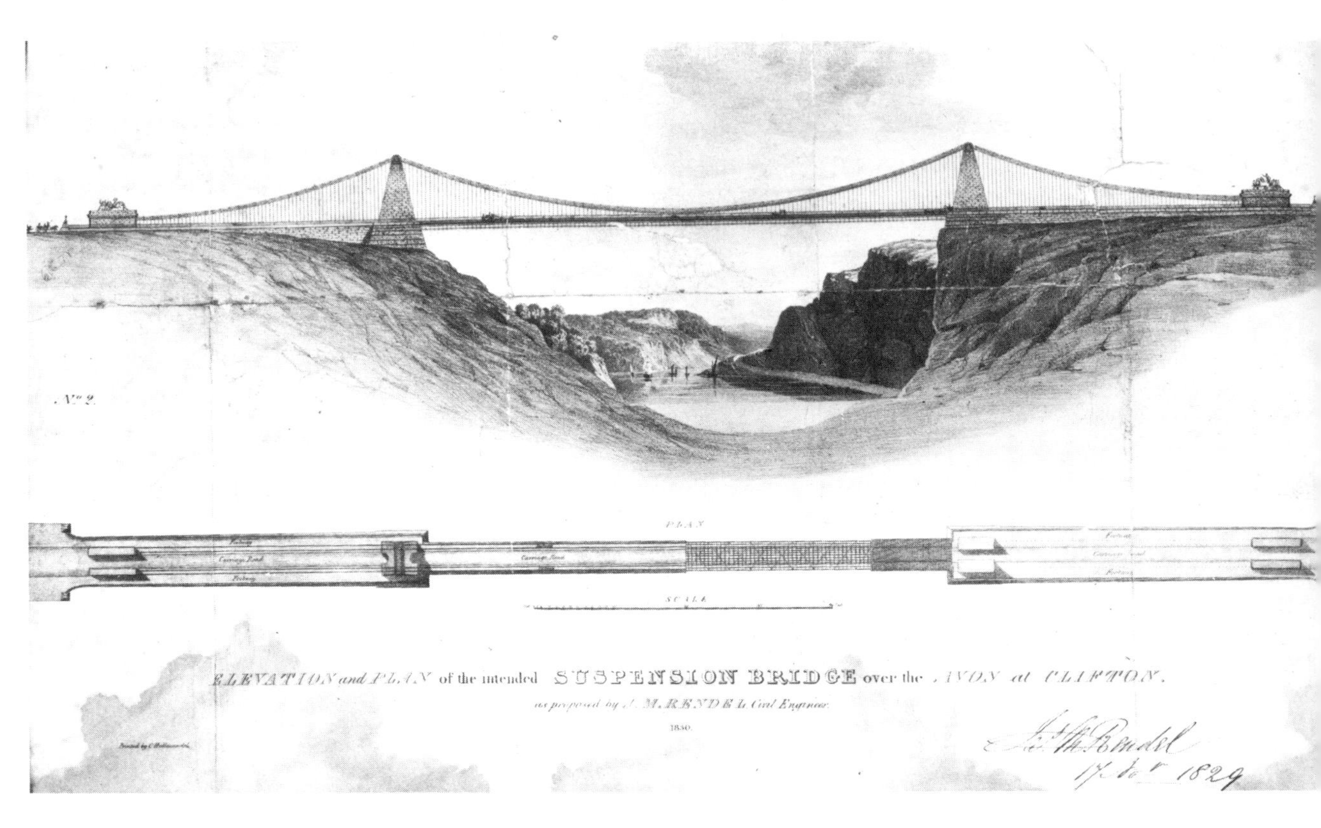

JMR's plan of his proposed suspension bridge across the River Avon at Clifton 1830.

Later he proposed a less expensive bridge sited lower down the valley. The span was to be much less than that proposed in his original design and involved the use of cast iron, masonry and ashlar towers. The chains and roadway were as originally proposed and the total cost estimate was reduced to £29,200.

The judges upon this occasion were Richard Trevithick's friend Davies Gilbert, a Past President of the Royal Society, and John Seward, a marine engine builder of distinction and a partner in the Canal Ironworks at Millwall. They chose four of the twelve designs for final selection eliminating altogether Telford's re-submitted original design, which the great man had disdained to alter. The four successful candidates were Isambard Brunel, W. Hawks, Samuel Brown and James Rendel. Greatly to Brunel's chagrin Hawk's design was initially selected, but following his determined representations, the judges reversed the first two places. Thereafter, Brunel had the total support of the Bridge Committee, and indeed the majority of the people of Bristol. Rendel's design was awarded fourth place, the judges claiming that it was too expensive, although subsequent events indicate that this was a questionable judgement. For the third time JMR was denied the opportunity of building a suspension bridge.

For the third time JMR was denied the opportunity of building a suspension bridge.

The dreadful riots in Bristol in October 1831 which Brunel witnessed at close quarters virtually put an end to fund raising and the foundation stone was not laid until 1836. Within a year the contractor went bankrupt and work continued at a desultory pace until all the money raised by the Bridge Committee, amounting to £45,000, was exhausted. In 1853, exactly one hundred years after Vick's bequest, the chains and other secondary materials were sold for use on the Saltash Bridge. No further progess was made during the remainder of Brunel's life. Following his premature death in 1859, his former friends and colleagues at the Institution of Civil Engineers rallied together and formed a company to complete the bridge. The engineers in charge, W. H. Barlow and John Hawkshaw, adopted a roadway stiffening system indistinguishable from that recommended by JMR and subsequently used by him when rebuilding Captain Samuel Brown's ill-fated suspension bridge at Montrose.

* * *

One hundred and fifty years ago, an engineer's life was not concerned solely with surveying, designing, and co-ordinating schemes. Economic studies and the preparation of feasibility studies played an important part in their daily routine. In November 1829 JMR completed a comprehensive economic study and a preliminary survey for a proposed tramway from Perran Porth on the north Cornish coast to Truro in the south. He studied the mining, agricultural and general business needs of the district and recommended the most suitable route for a horse-drawn tramway. The principal need at this time was for a more economic means of transporting Perran sand, a valuable manure, to Truro for onward shipment. JMR paid particular attention to the fall of the line favouring the direction in which the heaviest traffic flowed. He recommended the use of inclined planes operated by stationary steam engines for hauling the ascending traffic, and horses were to provide the motive power on the level stretches. He gave comparative costings for a line with continuous gradual slopes, thereby elminating the need for inclined planes, but greatly increasing the number and the cost of earth works required. He presented detailed figures of operating costs, maintenance charges and depreciation, suggesting the freight charges necessary to make the scheme viable. He estimated the total cost of his proposals as £19,300, but so far as can be traced the scheme was never implemented.

By 1828, the last unfinished link in the South Devon coast road between Exeter and Plymouth via Teignmouth, Torquay, Dartmouth and Modbury remained the crossing of the River Dart. Working under the aegis of his mentor, the Earl of Morley, in conjunction with the eminent Plymouth architect, John Foulston (1772-1842), JMR prepared plans for a suspension bridge across the river at Greenway Narrows, near the village of Dittisham. The promoters soon ran into difficulties following opposition led by James Elton, the owner of Greenway House and proprietor of the ferry at Greenway, and the scheme had to be abandoned.

Undaunted, JMR set about finding an acceptable solution to the impasse. Initially, he prepared plans for an endless chain ferry worked by manually operated machinery from the shore. During a visit to London he discussed his proposals with his friend and fellow member of the Institution of Civil Engineers, Henry Maudslay. By coincidence, Maudslay had recently engaged a young Scotsman named James Nasmyth whom, as has been already noted, JMR met whilst he was employed by Telford in Scotland. Apparently, Nasmyth had shown Maudslay his drawings of the proposed steam tug boat for the Forth & Clyde Canal during his interview, and Maudslay immediately suggested to JMR that his new employee's ideas might provide an excellent solution for crossing the Dart. We do not know if the two men actually discussed the scheme face to face, but within four months of Nasmyth joining Maudslay, JMR had drawn up plans for a unique ferry working in a similar manner to Nasmyth's steam tugboat, which he described as a 'Floating Bridge'. At no time did JMR lay claim to the principle of haulage along a chain, but confined his claim to the extension and application of Nasmyth's earlier proposals. In a subsequent paper given before the Institution of Civil Engineers he referred to the Floating Bridge as, 'new in all the arrangements of its detail, if not in principle, and professing to be a valuable addition to the existing means of promoting internal communication'. In any event, Nasmyth appears to have been entirely happy with the situation for, many years later, he recalled having seen the third of JMR's Floating Bridges at work on the River Tamar between Torpoint and Devonport:

Within four months JMR had drawn up plans for a unique ferry which he described as a 'Floating Bridge'.

> I had the pleasure to see this simple mode of moving vessels along a definite course in most successful action at the ferry across the Hamoaze at Devonport, in which my system of employing the power of a steam engine on board the ferry boat, to warp its way along a submerged chain lying along the bottom of the channel from side to side of the ferry was most ably carried out by my late excellent friend James Meadows Rendel Esq.

Colonel Sir John Seale, one of the principal promoters of the Dartmouth Floating Bridge Company in 1829.

JMR's plan for an 'alternative bridge' was exhibited before a crowded public meeting held at the Castle Hotel, Dartmouth on 3rd October 1829, only five days before the famous Rainhill Trials when Stephenson's 'Rocket' decisively won the battle for the locomotive engine, ending all talk of horses and fixed haulage engines. JMR's project was unanimously approved. The principal promoters of the new company, styled Dartmouth Floating Bridge Company, were the Earl of Morley, Colonel John (later Sir John) Seale, George Cary and Henry Woollcombe, and steps were taken immediately to obtain the necessary Act of Parliament. Later, they elected to the Committee of Management the Plymouth architect John Foulston who supervised the design of the piers and other buildings associated with the project. Initially, the Bill was opposed by the Borough of Dartmouth and by the Duchy of Cornwall, both of whom considered the ferry would interfere with their rights. However, when the Admiralty announced that they did not consider the scheme a threat to navigation the opposition collapsed and the Bill received the Royal Assent on 17th June 1830.

The site finally chosen near the Royal Naval College had a crossing distance of 1650 ft. The ferry comprised a twin-hulled vessel 40 ft long by 30 ft wide, with its 4 h.p. steam engine, boilers and hauling gear mounted between the two hulls and covered by an awning. The outer sections of the vessel, each 12 ft wide, were arranged to carry carriages, pedestrians and cattle. Two iron chains were employed for safety in case of breakage, and were securely attached to rocks on each side of the river, such that they would normally lie on the river bed and yet be sufficiently taut to maintain the direction of the ferry against the effects of wind and tide. The hull was built by the Plymouth boatbuilder, Isaac Blackburn of Turnchapel, who during the war against the French had built the two largest 74 gun warships ever built, outside the Naval Dockyard. The 'Floating Bridge's' engines and machinery were built by the Plymouth ironfounder, John Mare, who later, as a partner in Ditchburn & Mare of Millwall, played a major rôle in the construction of the rectangular iron tubes for Robert Stephenson's celebrated Menai Tubular Bridge. The ferry required three men to operate it, an engineman, a fireman and the toll collector, and the weekly running expenses were said to be little more than £5.

The 'Floating Bridge' was opened to the general public at mid-day on 19th August 1831. To the accompaniment of bands and the local militia, a great throng of people made their way down to the river along the new 1 mile long road to Lower Sandquay on the Dartmouth side. Some 60 carriages and 200 saddle horses then proceeded to Colonel Seale's home at Mount Boone Manor, where a sumptuous breakfast awaited them.

JMR's plan for an 'alternative bridge' was exhibited at the Castle Hotel, Dartmouth on 3rd October 1829.

Although the 'Floating Bridge' performed satisfactorily, the initial cost of the ferry and its approach roads, piers and quays greatly exceeded the amount the Company was empowered to raise under its Act. Extra capital was obtained by way of a mortgage, but never at any time were the shareholders paid a dividend. The situation was further aggravated by an extraordinary oversight on the part of the promoters. They had allowed a clause to be inserted in the Bill which exempted all Post Office mails from payment of tolls. A long struggle with the Post Office, led by Lord Morley, ensued, but to no avail. In order to comply with the law and the requirements of the Post Office mails, the Floating Bridge was obliged to operate a service for 18 hours out of every 24 irrespective of whether there was other traffic available or not. In 1835, the Treasury and the Postmaster General partially relented and agreed to pay the company £50 per annum compensation for having to operate the ferry when little other traffic was available. Unfortunately, the compensation was too little and came too late and in 1836 the steam engine was replaced by two blind horses working a treadmill amidships. Little imagination is required to anticipate the public outcry this solution would have caused in the present day.

JMR's own judgement on this sad state of affairs may be gleaned from a subsequent paper given before the Institution of Civil Engineers:

> From want of experience, which is found almost always to limit the success of first efforts involving many new arrangements and adaptations of machinery, the success of the Dartmouth Floating Bridge was but partial.

The end finally came in September 1855 when, during a violent storm, the ferry sank at its moorings. Nevertheless, the importance of Rendel's work was not overlooked and between 1833 and 1840 he was commissioned to construct 'Floating Bridges' at Saltash and Torpoint near Plymouth, and at Southampton and Portsmouth. Perhaps the most ambitious scheme of all was Robert Stephenson's railway ferry across the Nile in Egypt, which he evolved after a careful study of JMR's earlier arrangements. This extraordinary structure, looking more like a modern oil rig than a ferry, enabled trains to cross the river Nile mid-way between Cairo and Alexandria. The vessel's twin railway tracks were arranged such that they could be raised and lowered in order to align with the tracks onshore, irrespective of the level of the river.

Between 1833 and 1840 JMR was commissioned to construct 'Floating Bridges' at Saltash and Torpoint near Plymouth, and at Southampton and Portsmouth.

* * *

JMR was a communicator and extrovert and in no sense a lone wolf working single handed. He willingly conceded that he had much to learn from others, and appreciated the great benefits derived from his early association with the distinguished members of Lord Morley's Laira Bridge Technical Committee. He took every possible opportunity of discussing his ideas and problems with other members of his profession and participated in meetings of the Institution of Civil Engineers as often as possible. Mention has already been made of JMR's friend and brother-in-law, Samuel Elliott, whose father, John Elliott was a witness at his father's wedding in 1794. Samuel, a competent surveyor and draughtsman, undertook much of JMR's early road survey work and the preparation of his reports and drawings. JMR especially valued his close relationship with the distinguished architects John Foulston, his partner George Wightwick, and with Daniel Asher Alexander. Foulston had been a pupil of Thomas Hardwick, and was responsible for many fine Georgian buildings in Plymouth, including the prize winning Royal Hotel, Assembly Rooms and Theatre Royal. Alexander, also a Devonian, was architect to the Prince of Wales and Trinity House, and the designer of Dartmoor Prison. Prior to moving to Plymouth in 1829 and the establishment of the partnership with Foulston, George Wightwick (1802-1872) had worked for Sir John Soane in London as secretary-companion. He designed Plymouth Town Hall, Helston Guildhall and Grammar School and a number of churches in Devon and Cornwall. He was the author of many books on architecture and the theatre.

George Clarisse Dobson (1801-1874), a qualified civil engineer who married JMR's wife's sister, Augusta Harris, played an increasingly important role in the business following the inauguration of the

Bowcombe Bridge on the Kingsbridge road. The site of JMR's first hydraulically operated drawbridge.

Dartmouth ferry. His grandfather, a Norwich ironfounder, had emigrated to France where his family lived until the Revolution and as a young man George was taught to speak fluent French and Italian by his father. Formerly employed by the Cornish entrepreneur Joseph Austen, his first major assignment in JMR's employment was the preparation of the survey and plans for the Torpoint ferry completed in 1834. His son, Austen Dobson, born in 1840, married the daughter of Nathaniel Beardmore, JMR's first pupil. In later life Austen Dobson became the Poet Laureate, and his son Christopher was for many years Librarian at the House of Lords.

Beardmore was born in Nottingham in 1816, although at an early age he removed to Plymouth where his father took up a new appointment. Always regarded as a bright, intelligent boy, he was educated at a day school at Chudleigh and at Devonport Grammar School, where he lived with the Head Master, the Revd Henry Greaves. He began his professional training at the age of 15 as a pupil to George Wightwick, where he received instruction in draughtsmanship. Shortly afterwards he became JMR's first pupil and quickly acquired a thorough knowledge of surveying and the rudiments of civil engineering. Whilst still a pupil he was entrusted with the plans for a new road to Kingsbridge and assisted with important Admiralty work at Devonport and in Plymouth Sound.

The Kingsbridge road scheme included the building of an interesting hydraulically operated drawbridge at Bowcombe Creek near Kingsbridge in which JMR himself took a great personal interest. It was his first foray into the realms of hydraulic engineering which later played such an important rôle in his great works at Grimsby and elsewhere. We cannot do better than quoting from George Clarisse Dobson's report on the bridge given in the Proceedings of the Institution of Civil Engineers in 1843.

> This drawbridge spans one of five openings in a stone bridge, built across a navigable branch of Salcombe Harbour; it is in one leaf 15 ft 9 ins wide and 32 ft long, supported upon a cast iron shaft or axle, placed 7 ft 6 ins from the inner end, working in the abutment pier, which is built hollow to receive it, and thus the part within the axle-end acts as a counter weight. To the centre of the end cross-beam of the counter part, a chain is attached and after passing over cast iron sheaves in the masonry face of the abutment, is coiled on a drum fixed on a horizontal shaft, carrying on one end a pinion, worked by a rack, attached to the piston of the hydraulic press; by this means, motion is given to the shaft and drum, and consequently to the leaf of the bridge. Balance-boxes are hung to the counter-end by which means the shutting is regulated. The struts for supporting the leaf, when raised, are also thrown in and out of their places by a rack and pinion.
>
> The hydraulic press, used for opening and closing the bridge, is simple in its construction, and the whole works so easily that a female can open and close the bridge in about 15 minutes without difficulty. The fresh water used for the pump is returned into the reservoir every time after being used. The bridge was designed and erected by Mr J. M. Rendel, when he was engaged in improving the turnpike road in the south of Devon, about 1831.

In 1845, the bridge, now called Charleton Bridge, was replaced by a swing bridge. The weight of the moving section was carried on twelve cannon balls which were allowed to move freely in two grooved cast iron rings, one fixed to the movable roadway, and the other to the solid masonry below.

Beardmore's report still exists as a classic example of an early railway survey.

In 1838, upon the expiration of his pupilage, Beardmore took an office in London, and for a few months was engaged upon survey work in Surrey, and the extension of the London to Brighton Railway to Portsmouth. Later in the same year he was re-engaged by JMR, who had himself decided to move to London. He offered Beardmore a Partnership and left him in charge of the Plymouth office, where he undertook the survey of a proposed railway from Devonport and Plymouth, through Dartmouth Forest, to Exeter, with a branch line to Tavistock. Although superseded by Brunel's scheme for South Devon, Beardmore's report, published by JMR in 1840, still exists as a classic example of an early railway survey.

Other important works which occupied Beardmore at this time were improvements to the Devonport Water Company's service reservoirs and distribution mains supplying both the town and the dockyard. He supervised the building of Millbay Pier in a small tidal basin at the top of Millbay, Plymouth and various other river works for the Admiralty and other Government departments. The improvement of rivers and drainage became his main interest and his book, *Manual of Hydrology,* became a standard work. He safeguarded well JMR's interests in the Plymouth area until it was decided in 1843 that he should join his Senior Partner in London.

During his schooldays at Devonport Grammar School, Nathaniel Beardmore had two inseparable friends, Charles Greaves, a nephew of the headmaster, the Revd Henry Greaves, and John Coode. All three boys were born within a few months of each other in 1816, Greaves at Amwell in Hertfordshire and Coode at Bodmin in Cornwall. In Devonport they enjoyed access to a small workshop with a lathe and they spent all their leisure hours building model bridges and astronomical instruments. Greaves had bridges in his blood, his maternal grandfather, Robert Mylne, having designed the old Blackfriars Bridge over the Thames. It is a measure of JMR's well established reputation that, in 1831, Greaves's uncle, William Mylne C.Eng., selected him to train his nephew. Upon the completion of his apprenticeship in 1837 young Greaves joined his uncle as planned, but within a year he returned to JMR. He had already gained invaluable experience in the design and construction of 'Floating Bridges', and was immediately given the task of surveying the River Severn at Newnham. The Post Office Surveyor and Superintendent of Mail Coaches had recently reported to a Select Committee of the House of Commons on the need to improve the roads of Southern England to Milford in Pembrokeshire, to connect with the Irish Packet Boat. He had spoken favourably about the suitability of 'Floating Bridges' for crossing tidal estuaries and Newnham was selected as the best location for crossing the Severn. Probably, because of the rapid progress of railways throughout the Kingdom, the scheme was deferred and eventually abandoned. However, Greaves continued for many years to render JMR invaluable service in this field, both in the United Kingdom and later in India.

The third member of the trio, John Coode, became a most distinguished civil engineer. In 1847, when the Admiralty adopted JMR's plan for the formation of the Portland Harbour of Refuge he was appointed Resident Engineer. Later, following JMR's premature death, he became Engineer-in-chief and upon the completion of the works in 1872, he received a Knighthood; an honour that would almost certainly have gone to JMR had his life been spared.

He assisted JMR as a member of the Royal Commission on Metropolitan Sewage Discharge and sat on the International Commission of the Suez Canal. Following JMR's death he was appointed to the Royal Commission on Harbours of Refuge and became responsible for many of the refuges in the United Kingdom. At Bridlington he improved the harbour and the North Pier. He reported on the suitability of Filey Bay in Yorkshire as a harbour of refuge. He was retained by the Waterford Harbour Commissioners in Ireland and carried out major works improving the River Suir. In 1864 he was consulted by the Isle of Man authorities and during the next 20 years built the breakwater at Port Erin, the Victoria and Battery piers at Douglas, and improved the harbours at Port St Mary, Peel and at Ramsey. In 1867, at the invitation of the States of Jersey he designed the harbour at St Helier. In 1885 he was appointed Consulting Engineer to the Mersey Conservancy and became Chief Engineer of the harbour of refuge at Peterhead in Scotland. In 1890, he designed the new outer harbour for the Dover Harbour Board, but although an Act was obtained, he did not live to see the work completed.

Immediately prior to his death in 1856, JMR had been retained to advise the Cape Colony in South Africa on several harbour schemes around the coast. Coode continued this work and by 1876 was able to command a fee of £30,000 plus expenses for a single visit to the colony. He also undertook similar projects in Australia, New Zealand and Ceylon.

But, we are running ahead of our story, and must return to events following the opening of the Dartmouth Floating Bridge and the completion of the Exeter to Plymouth coast road. One suspects that perhaps JMR was in the doldrums during 1830. Although his small office was busy with various comparatively small local projects, he had no major schemes in hand, and faced the future with some trepidation. During the year he applied unsuccessfully for the post of General Surveyor to the County of Somerset, but perhaps not surprisingly, the job went to McAdam's son, James Nicoll McAdam.

One suspects that perhaps JMR was in the doldrums during 1830.

* * *

Shortly before the completion of the Laira Bridge, JMR was approached by the proprietors of the recently formed Torpoint Steam Ferry Company headed by the Earl of St Germans and the Rt. Hon. Reginald Pole Carew. They invited him to design new piers for a steam operated ferry they were planning to introduce between Torpoint and Devonport,

Brunel's Royal Albert Bridge at Saltash with JMR's 'Floating Bridge' in the background crossing the river Tamar.

across the Hamoaze, the estuarine mouth of the River Tamar. Armed with a letter of recommendation from the Earl of Morley, JMR met the committee in November 1826, and within a matter of weeks completed his plans for the piers which enabled the ferry to load and unload whilst laying in the direction of the stream. His report also advocated extending the sea wall which formed a part of the Devonport fortifications, utilising locally available Board of Ordnance stone. The work was completed in the summer of 1829, and the 70 ft long ferry, *Jemima,* named after the countess of St Germans, was launched at Richard Hocking's Stonehouse boatyard on September 29th. Similar to the two Tay Ferry vessels, *Jemima* had twin-hulls and was powered by a 12 hp steam engine driving a single paddle-wheel. Unfortunately, she proved quite inadequate to deal with the strong tides encountered in the Hamoaze, and was soon quietly withdrawn from service. The old horse boats formerly used were re-introduced.

In the months preceding this fiasco a group of Cornish landowners formed a committee with the object of establishing yet another ferry across the Tamar, this time at Saltash, two miles or so up river from Torpoint. They appreciated the need to improve the road access to Callington and Liskeard, because the growing traffic from Plymouth was compelled to take the long circuitous route through Tavistock. The committee were obviously impressed by the apparent success of JMR's 'Floating Bridge' at Dartmouth and invited him to prepare plans for a similar facility across the Tamar. This time the scheme was unopposed by both the Duchy of Cornwall and the Admiralty, and the promoters lost no time in securing the necessary Act of Parliament. The chosen site for the ferry was adjacent to that selected by Brunel 20 years later for his famous Royal Albert Railway Bridge. Opened in February 1833, the ferry was similar in all respects to that at Dartmouth. The twin-hulled ferry was built by John Pope, who had recently taken over Isaac Blackburn's shipyard, and John Mare supplied the steam engine and machinery. Unfortunately, it too had a short life and was compelled to cease operations in 1836 due to lack of funds. During 1832, whilst the ferry was being built, JMR surveyed the new road from Saltash to Trerule Foot where it joined the Torpoint to Liskeard road.

Memorial erected to JMR by the Tamar Bridge & Torpoint Ferry Joint Committee in 1978.

As soon as it became known that the Saltash Ferry was about to become

Hake's watercolour of JMR's Torpoint 'Floating Bridge' crossing the River Tamar — circa 1850.

a reality, the original promoters of the Torpoint Ferry re-activated their scheme and invited JMR to design an improved 'Floating Bridge' to replace the defunct *Jemima.* Their objective was to secure a crossing of the Tamar as a preliminary step towards the long cherished idea of a coast road from Plymouth to Looe, Fowey, St Austell and Truro. Quite independently JMR had already unsuccessfully tried to enlist support for such a road, and had gone so far as to make a new approach to Joseph Austen, putting behind him their earlier disagreement at Par Sands. But Austen remained cautious and was unwilling to participate in the road scheme until the 'Floating Bridge' was completed and had been given the opportunity of demonstrating its reliability during the winter months.

The river was 2,550 ft wide at Torpoint, having a maximum depth of 96 ft at high water. Built by Richard Hocking, the twin-hulled vessel was

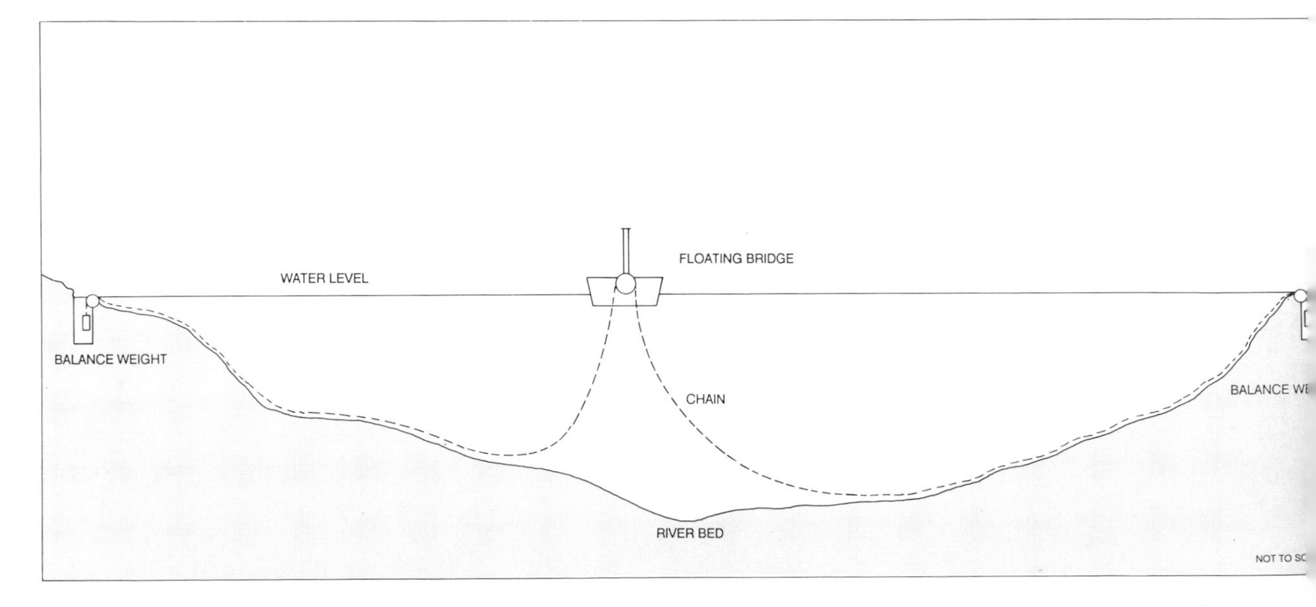

Diagram of JMR's 'Floating Bridge' clawing itself along a submerged chain.

of an entirely new design, 60 ft long by 50 ft wide and powered by two 11 hp steam engines built by John Mare. Variations in the chain lengths as the ferry traversed the river were continuously compensated by balance weights at both ends of the chains suspended in 20 ft deep shafts on the river banks. This ensured a smoother passage than that provided by the earlier designs and brought the vessel straight and steadily onto the landing ramps. The action of wind and current caused a lateral divergence of no more than 50 ft in mid-stream. The service was inaugurated in April 1834, and a second 'Floating Bridge' was introduced in 1835, the two

An early photograph of the Torpoint 'Floating Bridge' preparing to cross the Hamoaze.

vessels working alternate months. Both were replaced in 1871 and 1876 respectively.

During the second half of 1834, JMR employed a Cornish surveyor named Carveth, whom Austen knew and trusted, to survey the proposed road as far west as Cuddra Gate, near St Austell. Simultaneously, JMR concentrated on designing a very handsome suspension bridge across the river Fowey, on the site of the present car ferry from Fowey to

JMR's design for a proposed suspension bridge across the River Fowey, Cornwall – circa 1834.

Boddinnick. A surviving engraving indicates that he proposed a bridge about 500 ft long with 120 ft high towers at each end. The roadway was 80 ft above the high water mark. He proposed also a similar iron bridge at Looe, but no details of the design appear to have survived. JMR presented his report to the subscribers on 30th January 1835. He estimated the total cost of the scheme for a road from Torpoint to Cuddra Gate as £44,042, but expressly excluded the cost of the Fowey Suspension Bridge adding, 'I am instructed to omit the estimate (for the bridge), but to state that the requisite funds will be provided'. He emphasised that the route he had selected would reduce the distance between Devonport and Truro by 8 miles, representing a saving of 2 hours travelling time for a laden wagon.

JMR was to be disappointed and his fifth attempt to build a suspension bridge was frustrated.

But once again JMR was to be disappointed and his fifth attempt to build a suspension bridge was frustrated. By the time he presented his report several of the original subscribers had already had a change of heart and the St Austell to Lostwithiel and Lostwithiel to Liskeard Turnpike Trustees vigorously opposed the scheme. Even Austen became heartily sick of the wrangling. It had been hoped to present Parliament with revised plans at the beginning of 1836, but when these were completed after many delays, they were found to be so much at variance with JMR's proposals, Austen dared not show them to him. Eventually in August 1836, he agreed to pay JMR's fees in full, but not without adding that he considered them 'more than was fair or just!'

* * *

On 2nd September 1834 Thomas Telford, probably the greatest engineer England had ever known, died peacefully at his home in London. During the last seventeen years of his life, he devoted much of his time to the work of the Exchequer Loan Commissioners, for whom he acted as Consulting Engineer. A financial slump followed Napoleon's defeat at Waterloo, and it became extremely difficult to raise capital for new ventures. Many worthy projects were brought to a standstill and the Government of the day was concerned at the growing risk of serious unemployment. The Loan Commission was set up in 1817 to make capital available to deserving undertakings, and it became Telford's responsibility to assess the merit of all loan applications from an engineering standpoint. As a result there was scarcely a project of any consequence which did not at sometime or other claim his attention.

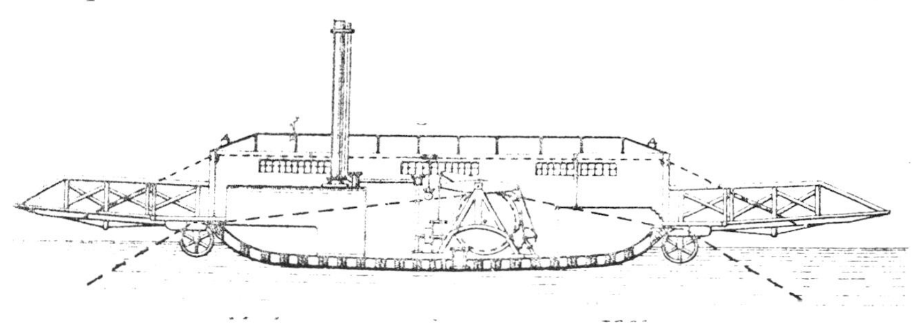

Torpoint 'Floating Bridge' showing disposition of beam engine and boiler.

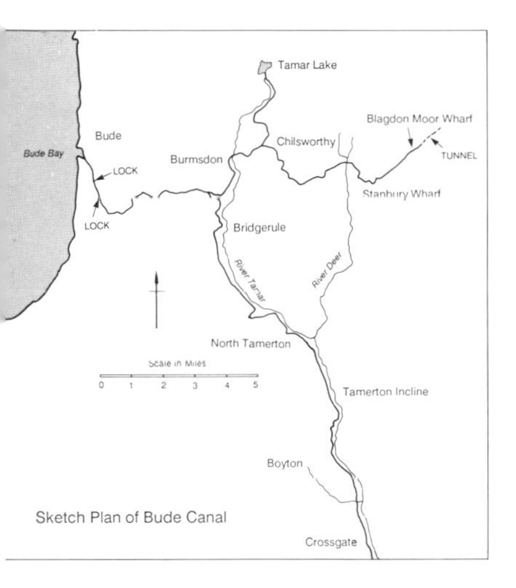

Sketch Plan of Bude Canal

In January 1833 Telford, acting upon behalf of the Exchequer Loan Commissioners requested JMR to report upon the state of affairs at the Bude Canal. Although first proposed in 1774, it was not until 1819 that Parliament sanctioned its construction supported by a loan from the Loan Commissioners. James Green, Lord Morley's former engineer, was responsible for both the design and construction of the canal which traversed the hilly country of North Cornwall and included no less than six unique inclined planes. The main traffic on the canal, as on the proposed Perran Porth to Truro tramway already referred to, was the local sea sand which was in great demand as a valuable manure. The specially designed tub-boats were provided with wheels for ascending and descending the inclined planes, five of which were worked by water-wheels and the sixth by a steam engine. JMR found the whole system in a very bad state of repair and was extremely critical of both the methods of working and the poor 'housekeeping'. However, his recommended remedies were expensive and in the end the proprietors preferred to take the advice of their engineer and JMR quickly faded out of the picture.

* * *

In the midst of building the Torpoint Ferry JMR was summoned to Southampton to meet the proprietors of the Itchen Bridge Company. Formed in 1833, the company originally intended building a swing bridge across the River Itchen near its confluence with Southampton Water. However, upon this occasion the Admiralty objected to the scheme, claiming that it would obstruct the free navigation of the river. They then took the unprecedented step of advising the promoters that they would not object to a 'Floating Bridge' and JMR was sent for with all speed. The necessary Act of Parliament was received on 25th July 1834 and orders were placed in Plymouth, with Richard Hocking for the twin-hulled vessel and with John Mare for the engines and machinery. The official opening ceremony took place on 23rd November 1836 attended by a large crowd of enthusiastic spectators. Unfortunately, within two years history repeated itself, and the high costs of maintenance and repairs began to exceed the toll income. Both Rendel and the contractors had difficulty obtaining their final payments, and although the company staggered on for over a decade, it was finally overcome by bankruptcy. The ferry ceased operations but after a visit to Plymouth in 1853 to see the new Saltash 'Floating Bridge', a similar vessel was established at Southampton in 1854. New ferry boats were again purchased at the end of the nineteenth century and these continued in service until replaced by modern, heavily subsidised diesel powered boats in 1967.

JMR's 'Floating Bridge' across the River Itchen at Southampton opened in November 1836.

In 1836, the Brixham Harbour Authority, then called the Quay Lords, commissioned JMR to prepare a scheme for increasing the capacity of the little Torbay port. Originally established in 1799, the local fishing fleet rapidly out-grew the capacity of the port's single exposed quay. JMR recommended that an entirely new outer harbour protected by two long stone breakwaters should be built, and the necessary Act of Parliament was obtained in 1837. Lack of funds delayed the start until 1842, and although some 1400 ft of breakwater was subsequently built, the work was again brought to a halt by further financial problems which ended JMR's involvement with the port. The breakwater was constructed from block stone quarried at nearby Berry Head. Stone facing was applied to the deposited blockwork on the seaward side only and was topped by a roadway and retaining wall. Although yet another disappointment for JMR, the experience gained at Brixham was of the greatest value and put to good use in the great works he later undertook for the Admiralty at Portland and Holyhead.

JMR recommended that an entirely new outer harbour should be built.

One hundred years before JMR's birth, Looe Pool was accessible to the sea as a deep inlet of Mount's Bay terminating at Helston where it formed the estuary of the small River Cober. In Tudor times boats sailed up to Helston, but gradually a bar formed across the entrance to the river completely isolating the waters of the estuary from the sea, and a new port was built at Porthleven. After heavy rain the upper part of the Pool quickly filled with water and threatened to flood the lower part of Helston.

The Revd Canon Rogers, the Squire of the vast Penrose estate adjoining Looe Pool, retained JMR to prepare a report on the practicability of removing the bar and forming a harbour at the mouth of the estuary, thereby removing the threat of flooding and providing a commercially viable harbour for shipping. In a long report published in May 1837, JMR set out the essential features necessary to overcome the difficulty. First and foremost the constant drift of the beach across the mouth of the estuary had to be prevented. This drift was found to move in both an easterly and westerly direction according to the direction of the

A view of Brixham Harbour in South Devon showing JMR's proposed breakwater — circa 1842.

prevailing wind. JMR proposed to deal with this by the construction of piers on either side of the river entrance at right angles to the beach and projecting out into 10 and 12 ft of water at low water spring tides. He recommended that the entrance to the Pool be at the south-eastern end of the bar under Carminow Point as being 'the best location from an engineering and nautical point of view'.

Secondly, JMR identified the need to increase the quantity and power of the backwater. To achieve this he proposed the construction of a dam across Looe Pool at Pentire Point having five sluices and a lock for the passage of ships. This arrangement would have provided a backwater in excess of 170 acres and allowed vessels of 300 tons to reach Helston. Thirdly, he emphasised the importance of correctly proportioning the width of the harbour entrance, which in this instance he calculated should be 200 ft. In an interesting appendix to his report he gave a resume of the history of the use of concrete. He recommended this be used for the construction of the piers and dam, pointing out that Telford had successfully used it when building St Katherine's Docks in London more than a decade earlier. He estimated the total cost of the project as £118,532.

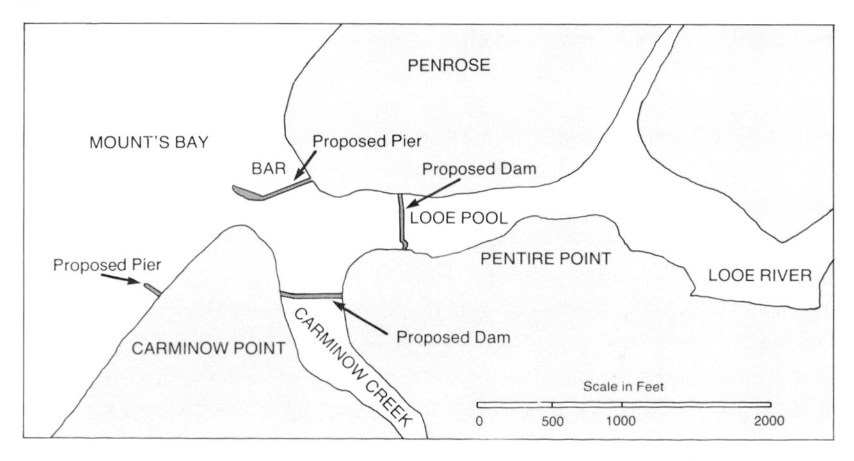

Rendel's scheme was never implemented, although later a tunnel was constructed under the bar in order to carry excess water from the Pool into the sea. In recent times both the Penrose estate and Looe Pool have been given to the National Trust and remain an area of outstanding natural beauty.

In September 1837 JMR was consulted by the Poole Corporation Harbour Committee who were having serious problems with the shifting bar at the entrance to their harbour. This was affecting navigation to Poole Quay and the channel had to be constantly dredged. JMR recommended that Lytchett Bay be made into a tidal reservoir linked by a navigable canal thereby increasing the backwater. He claimed that the cost of his scheme, which he estimated at £14,050, would soon be off-set by the savings in constantly having to dredge the channel. His report not only presented the Corporation with a highly professional appreciation of the techincal problems and their solution, but included proposals for a financial scheme designed to overcome the Corporation's embarrassing lack of funds. Once again nothing was done and the scheme lapsed. Shortly afterwards, in the early part of 1838, a few weeks before Queen Victoria's Coronation, JMR set up his office in Great George Street, Westminster.

In the early part of 1838, JMR set up his office in Great George Street, Westminster.

3

James Meadows Rendel
Civil Engineer, London 1838–1856

Rendel's decision to remove his office to London was prompted by a number of factors. Since his triumph at Laira he had had several disappointments. The 'Floating Bridges' had enjoyed only limited success and one suspects he regarded them as 'Alternative Bridges', temporary expedients until permanent structures could be built. Undoubtedly, he still had ambitions to build a major suspension bridge. The majority of the consultancy work he had undertaken for private companies and individuals had failed to materialise, due mainly to lack of finance. He believed that his most promising client in the future was likely to be the Admiralty with whom he had established important and cordial relations in London. The centre of gravity of his profession was London's Westminster district and he sensed that a partial vacuum had been created by the deaths of John Rennie senior and Thomas Telford. He saw the need to be on the spot when steering Bills through Parliament and appreciated that in future most of his committee and reporting work was likely to be in London. Furthermore, he thoroughly enjoyed his professional and social contacts with his peers and relished the idea of living amongst the technological elite of his profession who were established in London. He had developed close personal relationships with men like Robert Stephenson, Hawkshaw, Simpson and Fowler, and saw nothing but benefit being derived from the establishment of both his home and office at the very centre of the great works being projected. The British Empire and many of its neighbours were on the threshold of unprecedented developments. British finance was becoming more readily available for capital projects, both at home and abroad, and British brains and technology were in great demand. JMR was now ready to take his place amongst the leaders of his profession.

JMR was now ready to take his place amongst the leaders of his profession.

Initially, he put up his plate at 34, Great George Street, within minutes' walk of Parliament Square, Westminster Abbey and St James's Park. Later, he moved to No. 8. In both instances the family lived 'over the shop' and the affairs of the business not only occupied all JMR's waking hours, but completely dominated family life. Inspite of the constant comings and goings, many visitors have recorded that it was a happy home, always providing great warmth and hospitality. In later life, JMR's eldest Alexander, recalled the excitement of the family travelling by stagecoach from Plymouth to their new home in London shortly before the

Great George Street, Westminster, illustrating the Storey's Gate entrance to St. James's Park – circa 1840.

opening of Brunel's Great Western Railway. A near neighbour living at 24, Great George Street was Robert Stephenson and his young wife. Stephenson was then busily engaged upon the completion of the London to Birmingham Railway. JMR's second home at No. 8 is today the site of the splendid building housing the Institution of Civil Engineers.

JMR's second home is today the site of the building housing the Institution of Civil Engineers.

Concurrently with JMR's move to London, he was personally engaged upon two major works at opposite ends of the country. One was at Montrose, on the east coast of Scotland, where he had taken over responsibility from Telford for rebuilding Captain Samuel Brown's ill-fated suspension bridge. The other was at Portsmouth, where after a long, protracted struggle with the Admiralty, the 'go-ahead' had been given for the building of JMR's finest but final 'Floating Bridge'.

In December 1829 Captain Brown completed a 412 ft long suspension bridge across the river Esk at Montrose, replacing an unsatisfactory timber trestle bridge. Brown used the flat iron chain links and pins which he had patented in 1817 to form the two main suspension chains on each side of the bridge. The 10 ft long links were 5 ins wide by 1 in. thick and suspended in parallel curves from red sandstone towers. The roadway consisted of a flooring of 3 ins planks over which a 1¼ ins boarded floor was laid transversely and covered with a fine coating of sand and gravel bonded by tar. The planks rested on cast iron cross beams supported by 1¼ ins dia. suspender rods.

On 19th March 1830, a large crowd thronged on to the bridge to watch a boat race. As they rushed from one side of the roadway to the other, one of the chains failed and the majority of the people were thrown into the water with a great loss of life. The ensuing investigation was initially entrusted to Telford. Following his death in 1834, the Exchequer Loan Commission retained JMR to complete the investigation and supervise the repairs. He found that the links had been badly made and were bearing unevenly on the pins, bending and cutting into them. The rebuilding of the bridge, including the provision of additional suspension chains, was completed shortly before JMR moved to London. Unfortunately, in October 1838 a hurricane completely destroyed the new roadway, although the new chains and anchorages were undamaged.

MR conducted an nvestigation into the ehaviour of stiffening girders nd bridge decks when subject o high winds.

JMR immediately conducted a minute investigation into the behaviour of stiffening girders and bridge decks when subject to high winds. He established new guide lines for deep trussing the framing of bridge roadways in order to withstand the effects of undulation and oscillation caused by longitudinal wave motion and transverse sway. The memel planking was arranged diagonally layer upon layer and caulked. A 1 in. thickness of tar bonded sand and gravel was then laid on top of the roadway. The rebuilt bridge survived until 1930, when the greatly increased volume of heavy traffic in the district necessitated its replacement by a concrete cantilever structure.

During the rebuilding of the Montrose Bridge, Telford's Menai Bridge was twice severely damaged by hurricane force winds. Both carriageways were damaged and 444 suspenders were broken, causing a 175 ft section of the roadway to collapse, but the chains remained intact. Both William Provis and I.K. Brunel expressed conflicting views as to the cause and remedy, but JMR was quite categorical in his judgement. He maintained that the problems had arisen principally because too much emphasis had been placed on the theoretical properties of the catenary curve of the chains, and too little attention had been paid to the effects of the wind. He observed that the wind effect was rarely uniform, acting on the upper side of the roadway at one end of the bridge, whilst at the same time acting on the lower side at the other end. He objected to the use of stays and braces to counteract movements in the roadway, maintaining that the wind forces should be resisted by the design of the structure. In 1841 he presented a paper before the Institution of Civil Engineers describing his system for ensuring rigidity by vertical diagonal trussing above and below the bridge deck, which was received without dissent. When Brunel's Clifton Bridge was eventually built, a deck stiffening system indistinguishable from that used at Montrose 25 years earlier was employed.

JMR had been invited to submit proposals for a 'Floating Bridge' across Portsmouth Harbour with the object of giving direct access to Gosport, as early as 1834, the year the Torpoint Ferry was inaugurated. Admiral Sir Frederick Lewis Maitland, the Superintendent of the Portsmouth

Montrose Suspension Bridge after rebuilding by JMR in 1834.

Dockyard was instructed to hold an enquiry into the merits of the project. He was the officer who had accepted Napoleon Bonaparte's final unconditional surrender aboard HMS *Bellerophon* and transported him to Torbay in July 1815 following his defeat at Waterloo. His published findings were succinct and Wellingtonian in character.

> The navigation of the Harbour would be exposed to material injury from such an encumbrance and that it is therefore undeserving the countenance of Government. As the Harbour is the property of the Crown, the project may therefore be considered as at an end.

Nearly three years later, in October 1837, whilst at Montrose, JMR received notification from the promoters that the Admiralty had had a change of heart and had indicated that it was now in favour of the scheme. The following month JMR presented a detailed plan of Portsmouth Harbour, proposing not one, but two 'Floating Bridges'. Both were to operate from Gosport beach, one to Portsea, the other to a point in the old town near the Harbour entrance known as Portsmouth Point. Immediately, the Admiralty made it clear that they were unwilling to sanction two ferries, but left it to the promoters to make their selection. Following a lively public meeting, Portsmouth Point was finally chosen.

On 9th April 1838 JMR submitted his estimate of £15,000 for the 'Floating Bridge' and the works connected with it, offering to complete the project in eighteen months from commencement. By this date three-quarters of the money required had already been raised by public subscription and no time was lost in seeking Parliamentary approval. Inspite of opposition from the Watermen, who saw the project as a serious threat to their livelihood, the Bill received the Royal Assent on 9th May 1838. Events were indeed moving rapidly. The new company, styled The Company of the Proprietors of the Port of Portsmouth Floating Bridge, was registered and JMR was confirmed as the company's Engineer at a fee of £1,000.

The first successful iron ship had been built at Airdrie in 1819, but the conservative shipowners were slow to accept change. In the 1850's more than half the new vessels built in the United Kingdom still had wooden hulls. Naturally, informed people in Portsmouth, many dependent upon the sea and maritime trade for their livelihood, were well acquainted with developments in this field. The proprietors suggested to JMR that their new ferry should have an iron hull. Invitations to tender were sent out on 27th June 1838, but only D.E. & A. Acraman of Bristol met the specification, all other offers received stipulating wooden hulls. A Board Minute dated 25th August 1838 confirmed the award of the contract in the following terms.

Port of Portsmouth
Floating Bridge

I estimate the expense of constructing the proposed Floating Bridge and works connected therewith across the Harbour of Portsmouth, viz. from Portsmouth Point, to Gosport Beach, in the sum of Fifteen thousand pounds, and that the same may be completed in eighteen months from the time of commencement.

Jas. M. Rendel
Civil Engineer

34 Gt George St
Westminster
9 April 1838

JMR's written estimate for his Portsmouth 'Floating Bridge'.

> D.E. & A. Acraman of the Bristol Iron Works stand engaged for the construction of one iron bridge complete with a pair of steam engines with the requisite machinery and fittings and two chains of one and three-eighths inches diameter for the sum of £5,900, with the stipulation that they will supply two iron bridges complete with the requisite engines, machinery and

JMR's 'Floating Bridge' at Portsmouth.

chains (as above) for the sum of £11,350, provided the order of the same shall be given within twelve months after delivery of the first bridge.

Many problems had to be overcome during the following months. Nathaniel Beardmore was called in to supervise the drilling of the chain wells, bored out of solid rock on the Portsmouth side. Charles Greaves was sent to Bristol to supervise the construction of the 'Iron Bridge', which was the first iron vessel to be built at Bristol. The hull was 68 ft long by 60 ft wide on the waterline and the deck was 100 ft long. She was built in three sections at St Philip's and assembled on a temporary slipway adjacent to Totterdown Bridge prior to launching into the tidal river Avon on 25th September 1839. Although mourning the sudden death of his daughter, Catherine, JMR was present when the 113 ton Plymouth based paddle steamer *Sir Francis Drake* started her 300 mile journey towing the bridge to Portsmouth. She arrived safely ten days later on 7th December 1839. The two 17 hp beam engines were supplied separately and installed once the vessel arrived at Portsmouth.

The 'Iron Bridge' . . . was the first iron vessel to be built at Bristol.

The new service was inaugurated in JMR's absence on 4th May 1840, nearly six months behind schedule, a situation which clearly annoyed the directors, for a few days later they made the following statement, after imposing severe penalties on Acramans.

> The Bridge was opened by the directors on the 4th inst; but without any public ceremony for reasons which they hope will prove satisfactory to the company. In consequence of the delay which had taken place in completing the different works, from the period originally contemplated; it was an object of much anxiety to the directors that the bridge should be at work and earning money the very moment she was ready. . . . they accordingly abandoned all idea of any public procession, and they hope it will be considered, in so doing, they were consulting the best interests of the proprietors.

JMR's Portsmouth 'Floating Bridge' approaches the Gosport slipway.

A month later JMR attended the Admiralty in London accompanied by Admiral Sir Francis Austen, a director of the company and the brother of the famous novelist, Jane Austen. They secured the Admiralty's agreement to the purchase of a second bridge from Acraman, which was brought into service in October 1842. During the Great Exhibition in 1851 the two vessels were named *Victoria* and *Albert*. By 1859 *Victoria* had come to the end of her useful working life and was replaced three years later by a new craft, *Alexandra,* named after the Danish princess who became Edward, Prince of Wales's consort. *Albert* sank in a severe storm in the 1890's and *Alexandra* survived until 1959, bringing to an end the era of the Rendel 'Floating Bridge'. Following JMR's death in 1856, the engineer responsible for the Portsmouth 'Floating Bridge' was Francis Trevithick, formerly the Chief Mechanical Engineer of the London & North Western Railway and son of Richard Trevithick, the 'Father of both the high pressure steam engine and the railway locomotive'.

* * *

Within three years of establishing his office in London, JMR became involved in so many commissions it would be tedious to attempt to list them all. His solution to the problems at Montrose had greatly enhanced his prestige and reputation, although paradoxically a minority of his work was associated with bridges. The Admiralty and other Government departments, as well as local authorities and public bodies, sought his expert opinions mainly on matters connected with rivers, harbours, docks and other marine works. In every case he was retained in his professional capacity as a civil engineer, and never was he the promoter.

Admiral Sir Francis Austen, brother of the famous novelist, Jane Austen and a director of the Portsmouth company.

Before he left Portsmouth in 1840, he reported on the proposed new steam packet harbour and docks adjacent to the naval dockyard, and he designed a new dock and pier at Gosport. He spent some time at Kingston-upon-Hull where he submitted proposals for improving the existing port facilities and designed a new dock. For the first time he

e Portsmouth 'Floating Bridge' wing the disposition of the two lage chains and the hinged ding ramps.

became involved in the vexed problems of the Fenlands, problems that were to remain with him for the rest of his life. Since the Middle Ages millions of pounds have been spent in the never-ceasing battle to get the inland water safely across Fenland to the sea. In 1841 he proposed new harbour works at Polperro in Cornwall, and at Bristol he prepared plans for the new river walls on the banks of the Avon. The following year his old friend Lord Morley sought his assistance to restore the Chelson Meadows reclamations. James Green's simple flap type sluice doors had proved unsatisfactory, frequently becoming hinge-bound and clogged up, and most of the reclaimed land had once again become unusable. JMR solved the problem by designing new doors placing the centre of motion high above the centre of gravity of the doors.

On 23rd February 1843, JMR was accorded the honour of being elected a Fellow of the Royal Society. His application described him as a Civil Engineer, the inventor of the 'Floating Bridges' operating across the Hamoaze at Devonport and across Portsmouth Harbour, and the author of papers published in the Transactions of the Institution of Civil Engineers. Eighteen distinguished persons seconded the application recommending him as, 'deserving the honour and likely to become a useful and valuable member'. The list included several leading members of the engineering profession – James Walker, Sir William Cubitt, Sir Marc Brunel, William Mylne, General Pasley and George Rennie.

In the following weeks Nathaniel Beardmore joined his chief permanently at the London office, leaving juniors at Plymouth.

During the reconstruction of Montrose Bridge, JMR paid a nostalgic visit to Dundee and renewed his acquaintance with the proprietors of the Tay ferry. He found that although the original underpowered ferry boats had been replaced by bigger more powerful vessels, the company was still in serious financial difficulties and in danger of closure. The company had accumulated debts of £46,000 and received an annual income of only £4,900 of which £3,740 was required to cover the basic running costs, leaving insufficient money to carry out necessary repairs and maintenance and service the debt. He suggested that the Tay was an ideal place for a 'Floating Bridge' and quoted some interesting comparisons he had compiled regarding the operating costs of the Tay ferry boats and his 'Floating Bridges' in Southern England:

Tay Ferry	10s	2d per hour
Portsmouth	5s	0d per hour
Torpoint	3s	7d per hour
Southampton	2s	11d per hour

In August 1842 he submitted a long report containing his proposals for re-siting the piers and the provision of a 'Floating Bridge' at a cost of £11,380. He suggested that the cost may be met by the Exchequer Loan Commission. Although the proposals met considerable local opposition from vested interests and the ultra conservative Fife Commissioners, the scheme did get as far as a draft Parliamentary Bill.

In October 1843, JMR was called to give evidence before a Select Committee of the House of Commons. He found that his most severe critic on technical grounds was his recent seconder for membership of the Royal Society, Major General C. W. Pasley, H.M. Chief Inspector of Railways.

JMR was called to give evidence before a Select Committee of the House of Commons.

The General was concerned about the width of the river, which was 7500 ft compared with only 2550 ft at Torpoint. He believed this would greatly increase the weight of the suspended chain being carried by the vessel and that this would be detrimental to the ferry's performance and an added danger. JMR eloquently explained under cross-examination that the quantity of chain suspended was not in proportion to the width, but to the depth of the river, and that the Tay was half the depth of the Tamar at Torpoint. He pointed out that the strength of the tidal current on the Tay was substantially less than that found at Portsmouth. He further claimed that his 'Floating Bridge' would cross the Tay three times faster than the present ferry boats. The committee appeared to prefer the General's testimony and the opposition won the day.

At least the local press supported JMR for on 13th October 1843 the following editorial comment appeared in the Dundee & Perth Advertiser.

> The time has not long gone by since a railway engineer of the old school (a reference to George Stephenson's friend Nicholas Wood) – eminent in his own way, and like the Pasleys of the present day, of, 'confessed high authority' with minds of a biased and feeble cast – pronounced it impossible for railway speed to exceed 14 m.p.h. More recently Dr Dionysius Lardner in his wisdom limited the utmost extent of a steam voyage to a distance of 2000 miles . . . that Mr Rendel has reached the *ne plus ultra* speed on his aquatic railway is not more probable than was the result reckoned on by Mr Nicholas Wood in regard to railway conveyance. The Tay affords a fairer field for the genius and enterprise of this talented engineer than has ever before been presented in an undertaking of similar description.

One suspects that the events concerning the Tay ferry were of secondary importance in JMR's mind at this time, although they demonstrate his remarkable ability of being able to deal with a variety of matters of importance simultaneously. Some months earlier he had been approached by a group of Birkenhead businessmen led by the influential John Laird. His father William owned two steam packet companies and in 1821 established the Birkenhead Iron Works, which in time became the world famous shipbuilding company, Cammell Laird & Co. Ltd. The Laird family had proposed building docks at Birkenhead as early as 1827,

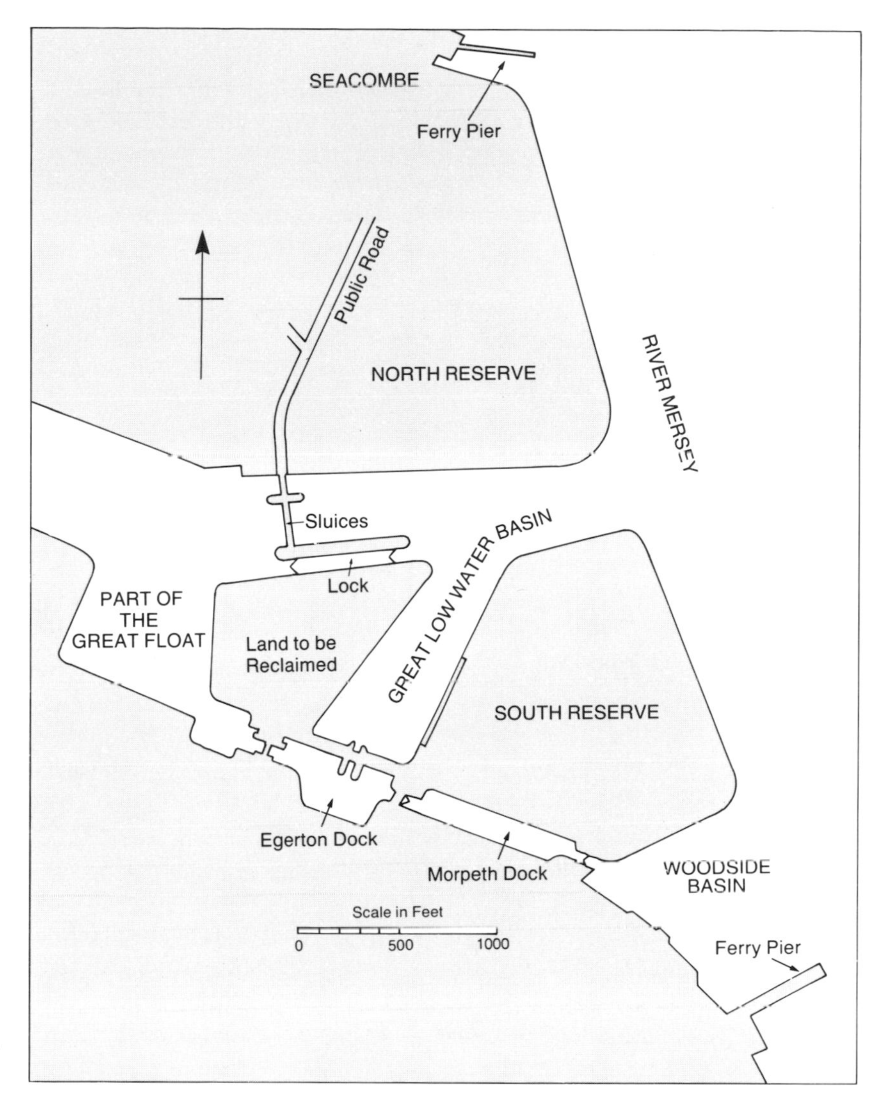

Diagrammatic illustration of JMR's proposed lay-out of the new docks at Birkenhead on the River Mersey.

but their scheme was frustrated by vested interests across the river in Liverpool. In 1843 the Birkenhead group proposed the conversion of Wallasey Pool, a creek off the Mersey, opposite Liverpool, into a floating dock and sought JMR's professional advice. On October 25th he presented his report and proposals.

The design of the docks brought JMR very prominently before the world.

Whether his original plans are ever completely executed or not, they will always remain remarkable in the annals of engineering for both the praise and condemnation which they have alternatively received. The design of the docks brought JMR very prominently before the world. In the years that followed he demonstrated his great ability defending his plans before numerous Committees of both Houses of Parliament against tremendous opposition. The collected evidence given by him and other leading engineers provides us with a valuable record of the state of engineering 140 years ago. The incessant mental anxiety inseparable from this undertaking was however more than even his powerful constitution could support indefinitely, and undoubtedly Birkenhead was responsible for shortening his life.

The site selected for the proposed dock at one time formed a deep low water lake situated between Seacombe Point in the west and Woodside in the east, and covered some 340 acres. The scheme provided an excellent opportunity for extending mercantile trade in the Mersey, whilst introducing competition for the established docks across the river at Liverpool.

Quoting from JMR's report

> I propose to construct a wall along the low water margin of the river Mersey, from Seacombe Point to the head of Woodside Ferry Pier. This wall will be parallel with the opposite defence wall of Liverpool docks. It will consequently have the effect of giving a truer current through this part of the harbour, probably deepening the channel on the Liverpool shore.
>
> About the middle of the wall I propose to leave an opening 300 ft wide, as an entrance to the basin of upwards of thirty-seven acres, excavated to a depth of twelve feet below low water spring tides, walled with convenient wharfs, and in every respect made suitable as a place of refuge for the mumerous vessels now obliged to seek shelter by running aground on the sand banks which constitute its site.

The report then dealt at length with the possible reduction of the Mersey backwater, which he did not consider detrimental to commercial interests on either side of the river. Perhaps rather provocatively, he pointed out that the new Chester to Birkenhead railway could quite easily be extended to the dockside giving passengers from the Irish Mail Packets then using Liverpool, easier access to London and other rail connected provincial cities. His report also proposed a 10 acre basin immediately adjacent to Woodside Ferry Pier to accommodate river and coastal craft which, formerly, had to discharge their cargo whilst lying aground on the open sand banks. This proposal, he pointed out, would also be of great convenience to the ferry steamers.

The promoters having accepted JMR's scheme launched their draft Bill on its passage through Parliament in June 1844, but it immediately ran into formidable opposition from Liverpool Corporation and other influential vested interests in the area. JMR was of course the main witness before the Committee of the House, and he was ably assisted by Sir William Cubitt (1785-1861) and George Bidder (1806-1878), a fellow Devonian whose powers of rapid calculation made him a fearsome opponent in committee. It is unfortunate that space precludes the reporting in detail of JMR's many remarkable performances before Parliamentary Committees. It was his ability and professionalism in these awesome tasks which enhanced his reputation as much as his acknowledged technical expertise as a civil engineer.

JMR was the main witness before the Committee of the House.

In due course, the Bill received the Royal Assent and the foundation stone was laid by Sir Philip Egerton on 23rd October 1844. In the following year the Birkenhead Dock Company was formally incorporated. But progress was extraordinarily slow and and the authorised capital of the company was expended with only a quarter of the work done. Soon the shareholders began to express dissatisfaction with their engineer and JMR was pressed to make alterations to the scheme which he adamantly maintained would cause only further delay and increased costs. Bad feelings and dissension continued for several years and efforts were made to appoint James Abernethy (1814-1896) as Resident Engineer. He had already made his name as Assistant Engineer at Goole Docks, Resident Engineer at Aberdeen and Engineer to George Hudson's North Midland Railway. In spite of his great reluctance to leave a job unfinished, JMR did not seek renewal of his contract when it expired in 1851.

But that was not the end of the matter. In consultation with Sir John Rennie, Abernethy produced revised plans in line with William Laird's original thinking and the promoters called upon Admiral Sir Francis Beaufort and Robert Stephenson to act as arbitrators. The hearing lasted from June to September 1851 and JMR was cross-examined on eight consecutive days. The Board's findings may be summarised as follows:

JMR's family home at 10 Kensington Palace Gardens, photographed in 1856.

> That having regard to the works which are already so far advanced, it would be highly injudicious to introduce such a radical change in the original project as that proposed by Abernethy, since all the evidence leads to the conclusion that the expense would be increased, and its usefulness seriously impaired.

The following year the promoters sought various minor amendments to the original Bill arising from suggestions made by the Arbitrators and the whole unhappy saga was once again publicly ventilated before a Parliamentary Committee. Again the outcome was a complete vindication of JMR, but he did not resume any active participation in the project. In 1855 the entire undertaking passed into the hands of Liverpool Corporation and three years later, in 1858, all Birkenhead docks came under the jurisdiction of the Mersey Docks and Harbour Board, which still exists as a company.

During a subsequent discussion at the Institution of Civil Engineers, a leading member, Mr Thomas Hawksley commented,

> All must be pleased that Mr Rendel's reputation has been entirely and completely redeemed from the imputation which might previously have existed with this work. It is well known that, with regard to the works carried on at Liverpool, between the two sides of the Mersey, there had been such strong antagonism, that it had almost been impossible for any engineer to do justice to himself, to his employer, or to the public at large.

James Meadows Rendel, F.R.S., President of the Institution of Civil Engineers in 1852/3.

The Institution of Civil Engineers awarded JMR its greatest accolade.

Indeed, during the very height of the controversy, the Institution of Civil Engineers awarded JMR its greatest accolade and elected him its President for the 1852/53 Session. He stood before the world, recognised as the head of his noble profession. His services were in great demand both at home and abroad, and he was able to command high fees for his services. That year he moved to a splendid house, 10 Kensington Palace Gardens, designed by his friend Thomas Henry Wyatt, the Honorary architect to the Institution of Civil Engineers.

But we are again running ahead of our story, for in the early stages of Birkenhead he undertook three other major projects – at Grimsby, Portland and Holyhead – which will forever rank as his greatest works.

Grimsby stood as a small town on the little River Freshney at the mouth of the Humber estuary. Its first enclosed dock was completed in 1801 by the Grimsby Haven Compar.y. In 1846, the recently formed and very ambitious Manchester, Sheffield and Lincolnshire Railway, formed by the amalgamation of three smaller railway companies and the Great Grimsby Dock Company, resolved to build a large port complex competing with nearby Kingston-upon-Hull. The eminent engineer, John Fowler (1817-1898), destined to be builder of London's first underground railway, was engineer to the M.S. & L. Rly. and JMR was appointed engineer responsible for the docks.

The proposed works, which went far out on the mud banks of the

Humber, requiring great skill in their execution, consisted of a basin 1,000 ft long by 700 ft wide having a 70 ft wide entrance, two locks, a dry dock, graving slips, repair yards, and 3500 ft long wharves complete with new road and rail links and a new passage to the old original dock. A 6,470 ft long embankment and cofferdam enclosed about 140 acres of reclaimed land.

The massive lock gates and the hydraulic machinery to operate them were designed jointly by JMR and William George Armstrong (1810-1900), about whom we shall learn more in the next chapter. The two men had become acquainted sometime during 1845, probably when JMR was first called in by the Conservators of the River Tyne in connection with the proposed North Shields Quay. JMR quickly saw the potential for Armstrong's ideas and encouraged him to establish an engineering works at Newcastle-upon-Tyne, which he styled The Newcastle Cranage Company, later changing the name to W.G. Armstrong and Company. Thereafter, Rendel and his family played a major part in Armstrong's business and private life. The water pressure required to operate the hydraulic machinery was supplied from a 300 ft high tower. This attractive structure was built in the style of the Italian Palazzo Publico in Siena and probably provided the inspiration for Carmichael's elegant painting of the dock.

The entrance to the new dock was designed to be beyond the limits of low water, allowing passage of merchantmen at all times. This necessitated embankments and an entirely self supporting cofferdam to be built out into the river ¾ mile beyond the old dock and 1 ½ miles in length in order to enclose the land to be reclaimed from the river. The cofferdam was subsequently described as 'a structure which has scarcely its parallel for magnitude, boldness of design and exposure of position'. Started in 1846, this part of the project was completed in 1848, and the first stone of the new dock was laid by the Prince Consort in the following year. The dock walls were also unique because of the soft, silty character of the back fill, and were formed as long piers with brick arching between, giving an unusual, but elegant appearance.

In 1852 the whole project was completed at a cost in excess of £1 million. Messrs Lynn of Liverpool built the cofferdam, and Messrs Hutchings, Brown & Wright executed the dock works, both performing well, and no major unforeseen technical problems were encountered. The promoters had at all times been co-operative and understanding, an altogether different story to Birkenhead.

The owners and builders were honoured by the visit of the Queen.

The owners and builders were honoured by the visit of the Queen and other members of the Royal Family to open formally the new docks on 14th October 1854. The Queen recorded the event in her journal:

> We then proceeded up the Humber, quite an arm of the sea, to Grimsby where we arrived before 1 o'clock. Albert laid the first stone of these fine docks in April 1849. There is an immensely high tower, 300 ft high, at the entrance, which is used for pressure by water and for the packing, lading and unlading of goods on the ships. There we landed, being received by Lord Yarborough (who was extremely confused) and the Mayor and Corporation, who presented an Address in an improvised room. Then, we walked to and up part of the tower, but I did not go to the top in a hoist, as did Albert and the children. After this we lunched hurriedly and entered the train . . .'

William Armstrong recorded the event with rather more exuberance:

> All went off very well at Grimsby except that the time was too short for the

JMR's Great Grimsby Royal Docks built in the period 1844 to 1853. Notable features were the 1500 ft long cofferdam used in construction and the 300 ft high hydraulic tower. Reproduced from an early coloured lithograph.

royal party to see much. Rendel escorted the Queen and Prince with the royal children over part of the works including the hydraulic tower up which Prince Albert, the Prince of Wales, the Princess Royal and younger ones with their tutor and governess were all hoisted by the water machine to the immense delight of the children. The Queen started laughing at them below. Rendel went up with them each time and I directed the working of the machine and saw and heard the whole affair. There was no time for any presentation or anything of that kind, but the whole thing was very amusing and certainly the hydraulics were a great attraction. Mr Rendel was in high spirits – the gates opened majestically and the machinery was in capital order.

★ ★ ★

Weymouth Bay, in Dorset, had for centuries been used as a harbour of refuge for all types of shipping. With the advent of the steam ship, the Admiralty realised that the adjacent Portland Roads protected by Chesil Beach were an ideal location for the development of a coaling and watering place situated mid-way between the great naval bases at Plymouth and Portsmouth. In 1844 a Royal Commission was set up and the Admiralty instructed JMR to prepare what today we would call a feasibility study. His scheme, estimated to cost £560,000, was accepted remarkably quickly and a start was made on the South Mole in 1847. JMR appointed his assistant, John Coode, as Resident Engineer, and once again the overworked Prince Consort was called upon to lay the first stone.

The great breakwaters were constructed entirely from locally quarried Portland Stone conveyed down to the sea on gravity operated inclined planes. The loaded wagons were arranged to cable haul the empty wagons

The first arm of Portland Harbour of refuge breakwater – circa 1860.

The workforce was provided by the 900 inmates of Portland Jail.

up to the quarry for refilling. Initially, the workforce was provided by the 900 inmates of Portland Jail, most of whom were awaiting transportation to Australia. The railway, supported on trestle work, was extended along the line of the proposed breakwater and the contents of the wagons were tipped into the sea to form a causeway. The 80 ft long piles supporting the trestles were shod with Mitchell's screw heads, which enabled the work to be performed with great speed and economy. The ashlar faced breakwater walls were then erected on this foundation. This stage of the work continued until sixteen years after JMR's death, and Edward, Prince of Wales, laid the last stone on 18 August 1872, when Coode received his knighthood. Further work was sanctioned in 1894 to give a total enclosed area of sea exceeding 2,200 acres, most of which was not less than 5 fathoms deep at low water.

In 1846, whilst engaged on the construction of Birkenhead docks, JMR was appointed Engineer to the recently incorporated Birkenhead, Lancashire & Cheshire Junction Railway. Always a somewhat wretched little company, very much dominated by the London & North Western Railway, it was formed to take over the original Chester to Birkenhead railway established in 1837, and to extend the line by 17 ½ miles from Chester to Walton Junction, near Warrington. The celebrated Cheshire born railway contractor and entrepreneur, Thomas Brassey (1805-1870), was appointed contractor upon JMR's recommendation.

During the remaining 10 years of his life JMR undertook several railway orientated commissions in the Merseyside and North Wales districts. Chester was emerging as an important railway junction with half a dozen companies seeking entry to the city, and JMR submitted plans for the new central railway station. These were rejected in favour of Francis Thompson's design based on his earlier work at Derby. The projected Chester and Holyhead Railway was in the Committee stage and JMR was instructed by the Admiralty to inspect and report on the proposed route of the railway between Conway and Bangor. George Stephenson was the Company's Engineer and he recommended a tunnel through the rock headland at Penmaenmawr, rather than going round it because of the danger from rough seas. JMR agreed with and supported this judgement.

The Admiralty then instructed Sir John Rennie and JMR to examine Robert Stephenson's proposals for the Britannia Railway Bridge across the Menai Straits, a few hundred yards west of Telford's Suspension Bridge. They made it clear that they would still not permit an arched bridge and insisted upon a clear headroom throughout of 100 ft. Stephenson was ready for them with his brilliant design for a gigantic tubular bridge. Originally, he had intended supporting the tubes by chains, but after discussions with George Bidder, John Laird, the shipbuilder, and William Fairbairn, these were discarded, leaving the tubes to be self-supporting; much to the consternation of Major General Pasley, whom we met at the Tay Ferry enquiry. Unlike his father, to whom a professional rival was necessarily a personal enemy, Robert was able to combine public rivalry with private friendship to a degree that set an example to his profession. This quality was also very much in evidence in JMR's character and he found no difficulty in expressing his professional judgement without causing offence to his colleagues or making enemies of his competitors.

The first substantial harbour works at Holyhead were carried out by John Rennie Senior, who constructed the Admiralty Pier used by the Irish Mail packet boats. With an ever increasing number of vessels using Holyhead as a harbour of refuge, and the appearance of larger ships in general, the Admiralty decided that further action was required. The passing of the Chester & Holyhead Railway Act further emphasised the need to improve the port's facilities as a packet station. Initially, James Walker and Captain Beechey were instructed to propose independently contingency plans, and in due course these were passed to JMR for assessment.

There can be no doubt that JMR was held in high regard by senior people at the Admiralty. Ever since putting up his plate in Plymouth he had made great efforts to cultivate relationships in Whitehall and at the principal Naval Dockyards, and Holyhead 'fell into his lap'. Without in anyway denigrating Walker's and Beechey's proposals, the bold alternative scheme he put forward was immediately accepted by the Admiralty, in spite of the cost being substantially in excess of his competitors' schemes. Of course it can be argued that JMR was able to demonstrate greater experience in this field and that his scheme showed greater vision and boldness. Nevertheless, knowing the right people seems to have paid handsome dividends upon this occasion.

The bold scheme he put forward was immediately accepted by the Admiralty.

The first part of JMR's scheme dealt with improvements to Rennie's Admiralty Pier. He proposed dredging and deepening an area of 12 acres to a depth of 12 ft at low water and providing three new timber jetties to accommodate the larger packet ships coming into service. Work was commenced in August 1846.

The main part of the scheme proposed a new harbour of refuge in the West Bay occupying an area of 316 acres and capable of holding several hundred ships. It was to have a northern and eastern breakwater and a steam packet pier 920 ft in length on to which the Chester & Holyhead Railway was to have direct access. After meetings with the railway company's Chairman, Captain C.R. Moorsom, JMR expected that they would meet the full cost of the pier, but that was not to be. After a period of intense railway politics, JMR embarrassingly had to admit that he had not provided for the cost of this pier in his estimates. The Admiralty were determined that the scheme should not fail and the Government's Holyhead Harbour Act and the Chester & Holyhead Railway, Holyhead Extension and Amendment Act both received the Royal Assent in July 1847. The latter Act provided that the railway should make a maximum contribution to the harbour project of £200,000, but

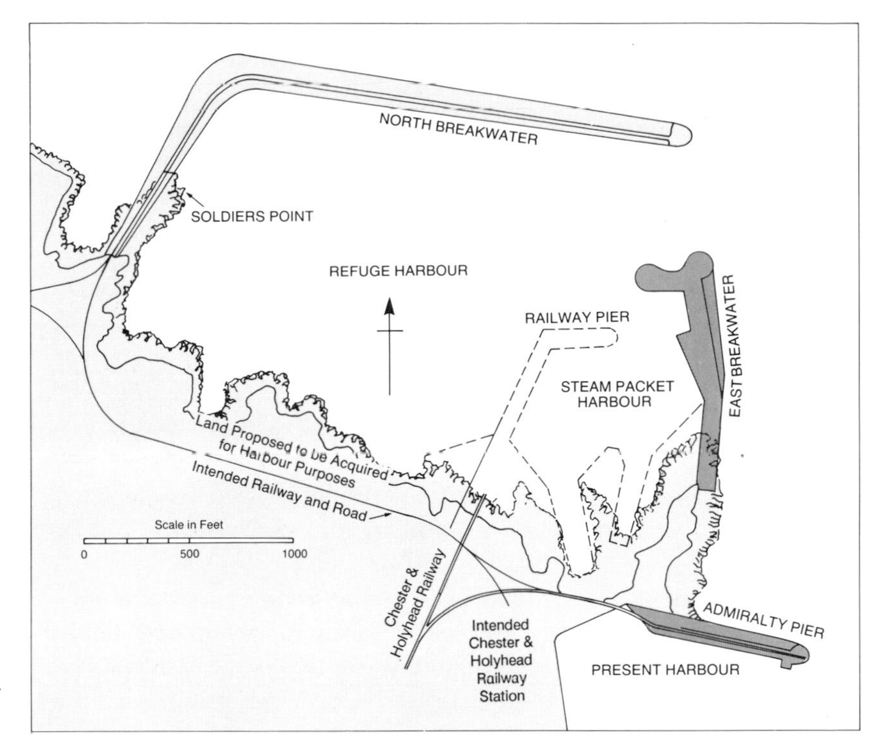

JMR's layout of the great Refuge Harbour at Holyhead on the Isle of Anglesey, North Wales.

within a year, events enabled them to withdraw completely from these works, leaving the Treasury, the Admiralty and JMR in control.

The work on the eastern breakwater and pier commenced in January 1848 with J & C Rigby of London as main contractors and JMR's brother-in-law, George Dobson, as the Resident Engineer. The first twelve months were occupied laying down a 7 ft gauge railway for bringing large stones from the nearby Moelfre and Holyhead mountain quarries, erecting the stages and making other necessary preparations for the works which already occupied over 1,000 men. Exactly the same methods of construction were employed as at Portland. During the winter of 1848-49, after several thousand tons of stone had been tipped along the line of the eastern breakwater, storms and heavy seas brought the works to a halt. JMR considered it unsafe for the men and stone trains to work on the staging, and decided to concentrate on the northern breakwater in order to give protection to the new packet steamer pier as quickly as possible. The construction of this breakwater presented even more formidable difficulties owing to the great depth of the water, which in places exceeded 55 ft. The foundations were built on a rubble mound, 250 ft to 400 ft wide at its base, and it was continually necessary to make good the huge blocks of rock swept away by the heavy seas before the sea wall could be consolidated.

Tragedy befell the family in 1851 with the death of Lewis Rendel.

Tragedy befell the family in 1851 with the death of Lewis Rendel, JMR's second son, aged 21. He was a talented young man with a scientific bent and was being tutored to succeed his father eventually. Working under George Dobson he had been associated with Holyhead since the work started.

By mid-1853 the northern breakwater had reached a length of nearly 5,000 ft, some 600,000 tons of stone having been deposited in the previous twelve months. Day after day the surrounding districts echoed to the noise of the huge blasting operations taking place on Holyhead Mountain, which were necessary to satisfy the project's apparently insatiable demand for stone.

Everybody was heartened by the unexpected visit of the Royal Family, who put into the harbour in the Royal Yacht during a violent storm. The

Queen recorded in her journal on 28th August, 'Albert went with Rendell (sic), the engineer, to see the Harbour or sort of Breakwater, inside which we were lying'.

Ships of every description began using the harbour while the work progressed, and upon occasions there was severe congestion, so much so that it was decided to extend the northern breakwater by a further 2,500 ft thereby increasing the capacity of the refuge to over 400 acres. Despite the extensive works already carried out on the eastern breakwater which extended 2,000 ft, there were continual changes of plan, forced largely by the changing requirements of the Chester & Holyhead Railway, and this part of the works was never completed.

Following JMR's death, Sir John Hawkshaw, a distinguished civil engineer with wide experience of both docks and railways, took over as Engineer and continued to supervise the works for a further seventeen years. Both George Dobson and Messrs Rigby & Co. were retained until the work was completed, and the harbour, then controlled by the London & North Western Railway, was formally opened by the Prince of Wales on 17th June 1873. Today, this great harbour has sadly declined in importance. It is still used by the Irish boats and by tugs and salvage vessels, but most of the great expanse of water within the breakwater is used as moorings by the yachting fraternity. A long jetty belonging to the Anglesey Aluminium Company now replaces the line chosen by JMR for the eastern breakwater.

Throughout the great works at Birkenhead, Grimsby, Portland and Holyhead, works which ensured JMR's name would be handed down for posterity besides those of Smeaton, Rennie and Telford, he and his growing band of assistants carried out a multitude of other assignments to which we can make little more than a passing reference. In 1843, JMR was commissioned to advise on harbour works at the Essex port of Harwich, which was becoming increasingly important as a packet steamer and passenger terminal serving the Northern European ports; and at Swansea, where there was a large increase in bulk cargoes such as coal, iron ore and copper. He spent some years reporting on the small south coast ports at Langstone, Chichester, Arundel, Littlehampton and Newhaven. His assistant, Charles Greaves produced a masterly survey of the coast from Selsey Bill to Portsmouth and formulated proposals for a ship canal to Bosham. Several land reclamation schemes were also proposed in the area, but these were never carried out, due mainly to the Admiralty's fears of their effect upon the approaches to Portsmouth Harbour.

Charles Greaves produced a masterly survey of the coast from Selsey Bill to Portsmouth.

In 1844, the Grand Junction Railway sought Parliamentary powers to build a railway crossing the Mersey at Runcorn Gap, and to support it on a viaduct founded in the river. Naturally, the Bridgewater Navigation Trustees, the successors the builders of Lord Bridgewater's great canal in the area, were concerned about the possible effects this would have upon their interests. The matter was brought to the notice of the Admiralty, who enlisted JMR's assistance. In 1845 he submitted proposals which appeared to satisfy all the parties interests, but nothing was done until 1863, when the London & North Western Railway built a massive viaduct, thereby considerably shortening the route from Crewe and the south to Liverpool.

The Lynn & Ely Railway, designed to connect with the Eastern Counties Railway at Ely to give rail access to Cambridge, Norwich and London, received Parliamentary sanction in 1845. The promoters wished to establish a 1½ mile long extension to Lynn Harbour as quickly as possible. As a result of the acute shortage of draughtsmen and surveyors due to the 'Railway Mania' producing a boom in the new railway construction all over the country, JMR was invited to submit proposals. Perhaps unwisely in the circumstances, he submitted a bold scheme for an 8 acre, 20 berth dock with suggestions for an integrated approach to the needs of river, road and railway. The estimated cost of £100,000 clearly frightened the promoters and JMR was quietly dropped.

The 1840's were a period of unprecedented maritime development in the United Kingdom.

The 1840's were a period of unprecedented maritime development in the United Kingdom and in 1845 the Government set up the Tidal Harbours Commission. They were concerned about the poor state of many of our harbours and the dangers caused by wrecks which littered many approaches. JMR declined the offer of membership, no doubt feeling that it would place limitations upon his business activities, although he did, on more than one occasion, give evidence before the Commission. During the same year he and his friend George Wightwick, were invited to become Directors of the recently established Great Western Docks Company, which formed a branch line terminus of the South Devon Railway to connect with the Mail Packet Steamers using Plymouth. The Company took over the works Beardmore had supervised shortly after JMR moved to London, but it appears to have been a short lived interlude, because all subsequent works were carried out by Brunel, who dominated totally the railway scene in the West Country.

Garston, six miles up river from Liverpool, was an ideal location for the establishment of a dock capable of handling bulk shipments from the St Helens and Wigan coalfield, as well as serving the growing industrial areas of Widnes and Warrington. In 1845, the dilapidated St Helens & Runcorn Gap mineral railway merged with the Sankey Brook Navigation and JMR was invited to advise them. The Directors' original plan was to build a new railway along the banks of the Sankey Canal, but he persuaded them to abandon the scheme and concentrate on an extension to Garston, where he proposed a new deep water dock should be built. Due to lack of finance it was never a very grand affair during JMR's lifetime, but after absorption by the London & North Western Railway in 1864, it became the finest railway-owned dock complex in the United Kingdom. In the early part of 1853, JMR was recalled following serious damage to the dock caused by high tides and storms during the previous December. He recommended that the broken fissures be replaced by masonry piers and that a 12 ft high by 3 ft thick protection wall be built on top of the existing river wall. He suggested also that a timber boom be constructed across the large entrance gates, which were operated by hydraulically powered capstans supplied by his friend William Armstrong.

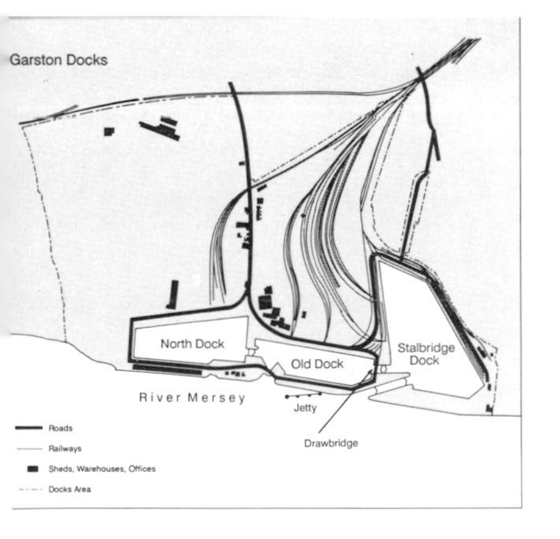

Within weeks of his first meeting with the St Helens & Runcorn Gap Railway Directors, JMR was instructed by the Treasury to report on various proposals put forward by Sir William Cubitt and James Walker for the improvement and extension of Leith docks near Edinburgh. A state of indecision had existed for many years and the local traders were becoming impatient. Early in 1846, he submitted a long report endorsing his professional colleagues' earlier proposals that a low water pier, an improved harbour channel, and a new dock suitable for the largest steamers should be built. He further advocated the extension of the Edinburgh & Leith Railway to the dockside and the building of a new roadway to the dock across a drawbridge. He costed his scheme at £135,000 and the necessary Bill went through Parliament without difficulty. The works, which proved of great benefit to the trade of the district, were completed in 1855.

At the time of his tragic death in Holyhead in 1851, JMR's son, Lewis, was engaged to be married to Emma Smith, eldest daughter of Charles Hamilton Smith of Plymouth. Smith was born in Flanders, educated in England, and trained as an engineer at the Austrian Engineering and

St. Peter Port Harbour, Guernsey – circa 1854 – showing JMR's method of constructing the breakwater.

Artillery School at Mechin. He joined the British Army and in 1816 fulfilled a secret mission in America, which earned him the rank of Colonel. He retired to Plymouth, where he became a great friend of the Rendel family. His name was amongst those listed in JMR's application for Fellowship of the Royal Society. In 1808, Smith had married Mary Anne Mauger, daughter of a prominent Guernsey resident, and it is undoubtedly this connection which paved the way for JMR's invitation to the island.

St Peter Port, the island's only harbour, had been in use since the reign of Edward I and had been enlarged by successive generations, but by 1840 further works became necessary to accommodate the new steam packet boats coming into service. JMR lost no time submitting his proposals and the scheme was approved by the Island States or Parliament, on 6th June 1851. Hutchings, Brown, Wright & Company, who did so well at Grimsby, were awarded the contract and work started almost immediately. Following delays occasioned by the Island's Lieutenant Governor, the contractors went out of business, and a Jersey firm, T. C. Le Gros, took over the work, but further interventions frustrated all attempts to get on with the job.

In 1854 the States realised that they were not doing justice to their Engineer's proposals and they requested he proceed, adding a low water landing place to the scheme. Although the new South Pier had been completed, the revised 1855 plan called for an alteration to the line, and an enlargement of the floating dock, both of which were approved by the States. At the time of JMR's death, the revised South Pier, the shipways, and the 650 ft long sea-wall and 250 ft long quay-wall had been completed. A start had also been made widening the west quay of the old harbour. He was succeeded by a Mr Lyster, as Engineer-in-Charge, but the works have never been completed, and the arguments continue.

In 1850 JMR had his first involvement with the great Port of London Docks.

In 1850, JMR carried out some consultancy work for the East and West India Docks, his first involvement with the great Port of London Docks built by Rennie and Telford. He persuaded the port authorities to install Armstrong's patent hydraulically operated cranes and hoists, greatly speeding the loading and unloading operations of ships using the docks. In 1851, he designed the new quay wall at North Shields on the Tyne, and in 1852 he put forward bold proposals for the establishment of new docks at Avonmouth, designed to overcome Bristol's growing problems associated with the seven miles of tidal river between the city and the sea. But nothing was done until 1877. He undertook a commission at Margate,

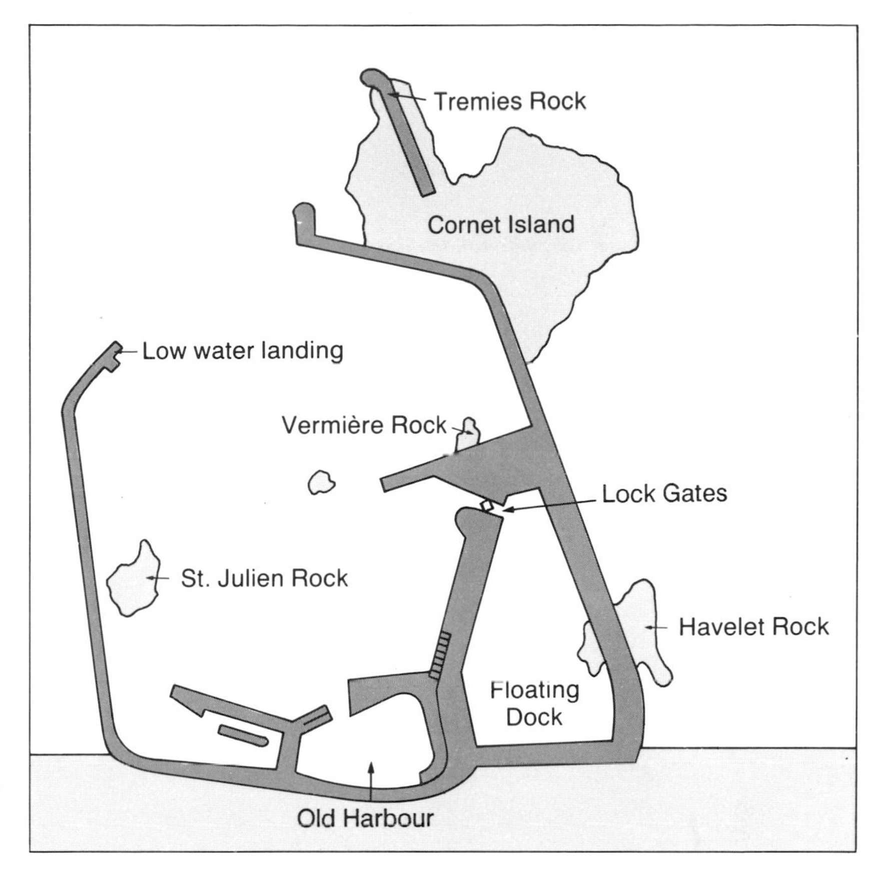

JMR's proposed layout of the new harbour at St Peter Port on the island of Guernsey — circa 1851.

where extensive new harbour works were proposed, which resulted in nothing more than a new 1240 ft long high water landing pier being built. In January 1878, violent storms lashed the east coast and the iron pier collapsed, never to be restored. Other commissions were undertaken at Sheerness on the Thames, at Whitehaven and Maryport in Cumbria, designed to support the growing iron industry in the area, and at Wick in the north of Scotland, which he described as the best natural deep-water harbour north of the Cromarty Firth.

JMR initiated some important experiments to determine the strength of Portland cement.

Portland cement was first produced at J. B. White & Sons works at Swanscombe, Kent in 1845. The properties claimed for this improved type of cement, which had its origins in Roman times, were of immediate interest to civil engineers and contractors. In 1847, JMR initiated some important experiments to determine the strength of this new material at Henry Grissell's Regent Park Ironworks in London. He tested to destruction specimens 9 sq ins in section by 18 ins long on a 75 ton hydraulic compression press. The results were fully recorded in the Proceedings of the Institution of Civil Engineers in 1852. Reinforced concrete, which we take so much for granted today, was not introduced until the 1880's.

In 1847, the Government set-up the Metropolitan Commission of Sewers to supersede the eight separate bodies hitherto responsible for sewers in allotted areas. Confirmed by the Metropolitan Sewers Act of 1848, the commission's members were Robert Stephenson, Sir William Cubitt, Thomas Hawksley, George Bidder and JMR, all at sometime Presidents of the Institution of Civil Engineers. Potentially, this development should have resulted in great improvements in what was becoming a national disgrace. Unfortunately, the Commission's first act was to abolish cesspools, discharging their contents into sewers which were allowed to find their way into the Thames tributaries, making the state of the river even worse. One hundred and sixteen schemes for main drainage were considered by the Commission during its comparatively short life, and in 1856 it was taken over by the Metropolitan Board of Works.

During 1847, JMR and Beardmore prepared plans for the complete reconstruction of Edinburgh's water supply, and the necessary Act was obtained shortly before the partnership was dissolved. New impounding reservoirs were built in the surrounding hills, a completely new system of distribution was installed and for the first time the inhabitants of the Scottish capital were assured a plentiful supply of clean drinking water. Six years later JMR undertook the same commission in the heavily industrialised City of Leeds. After leaving JMR in 1848, Beardmore concentrated on the River Lee navigation and other similar Thames drainage schemes, which he fulfilled with great distinction.

JMR and Beardmore prepared plans for the complete reconstruction of Edinburgh's water supply.

During the period 1852-3, JMR spent some time in Ireland accompanied by Sir William Cubitt. The Government were anxious to find ways and means of alleviating the plight of the Irish following the 'potato famine', which had resulted in starvation, death and mass emigration on a huge scale. The Treasury called in the two experts in the belief that large-scale land drainage in the centre of the country would enable unusable land to be brought under cultivation and provide desperately needed employment. They examined 385 miles of river and over 1 million acres of catchment basin and presented a very long report to the Treasury. When finally brought before Parliament and the estimated costs became known, the attitude of the majority was that the Irish should be responsible for their own salvation and nothing was done.

* * *

By this time the name Rendel was becoming known internationally, and although JMR himself did not travel abroad very much, his ambitious young assistants were only too pleased to have the opportunity of seeing the world and being able to tackle new and exciting problems overseas. In 1852, William Pole (1814-1901) joined JMR. Born in Birmingham and trained by the Horseley Company, he was appointed the first Professor of Engineering at Elphinstone College in Bombay before his 30th birthday. Ill-health forced his resignation, and he spent some time in Italy learning the language before returning home. Whilst in Italy he was introduced to senior Government officials concerned with the Port of Genoa and the scheme to remove the naval base from Genoa to La Spezia.

In March 1853, JMR visited Genoa accompanied by Pole, and met Count Cavour, President of the Council of Ministers, and the Marquis d'Azeglio. He found the systems employed in the Italian port archaic, taking up to 38 days to perform the work done in an English port in one day. He returned the following August with proposals for the modernisation and enlargement of the port, sufficient to meet its projected needs for the next 15-20 years. He proposed the building of new quays, a 7-storey warehouse, and the extensive use of Armstrong's hydraulic machinery for operating the dock gates and the dockside cranes. His proposals resulted in the development of a 140,000 sq metre site at a cost of over £1 million.

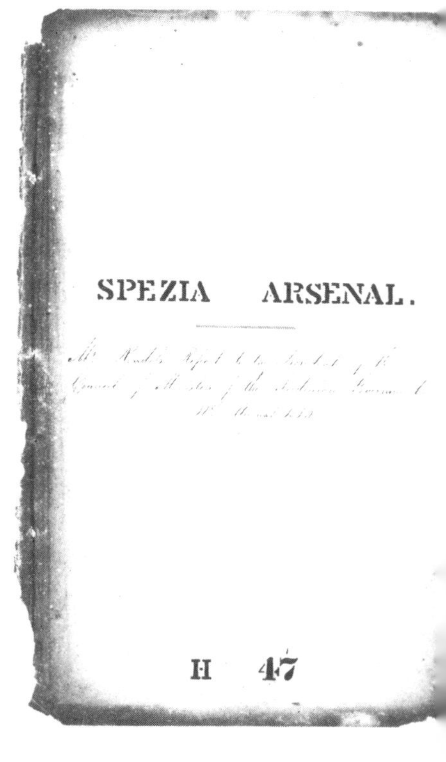
SPEZIA ARSENAL.

H 47

At Spezia, he submitted plans for the new Arsenal with its maze of interconnecting tunnels. His plans included the naval basin, also graving docks, slipways, and shore buildings, including workshops, barracks, officer's quarters, and a prison. He dealt also with road and rail connections, and an aqueduct to provide the base with an adequate water supply. Special provisions were made for the steampowered warships being introduced into the navy, including mechanical coaling facilities. Most naval bases throughout the world evolved over a long period, but at La Spezia, in the Bays of Le Grazie and Varignano, JMR built from scratch a large modern base such as had never been seen before. Pole stayed with

JMR built a large modern base such as had never been seen before.

Rendel until 1858 and then put up his plate in Storey's Gate, where he enjoyed a long and distinguished career.

In 1854, JMR was commissioned by the Admiralty to advise on a new harbour of refuge in Table Bay at Cape Town. He did not visit South Africa himself, but entrusted the survey work to John Coode, whilst ensuring that the final reports submitted in 1855 bore his signature. The breakwater was formed in exactly the same manner as Portland. After JMR's death, John Coode was appointed Engineer-in-Chief and the first load of rock was deposited by Prince Alfred, the Duke of Edinburgh in 1860. The project took 10 years to complete. Coode was responsible also for harbour works at Port Alfred on the Kowie River, mid-way between Port Elizabeth and East London, based on a scheme devised by JMR in 1854. He had laid down that the piers at the entrance to the river should be 250 ft apart, but for some reason the distance was reduced to 170 ft, a fatal error, causing great danger to shipping and contributing to the failure of the harbour. In some respects the scheme turned out to be another Guernsey, for completion was delayed until the end of the century. Another Rendel pupil, Charles Neate, who had been employed at Birkenhead and Grimsby was subsequently engaged by the Admiralty to do work at Port Elizabeth and Durban.

Assistants surveyed the 600 miles long railway from Madrid to Oviedo.

Upon the completion of the work at Grimsby in 1853, Charles Neate was sent out to Brazil for two years, where he surveyed new docks at Rio de Janeiro and then moved north to Pernambuco (Recife) to survey the proposed railway to Joao Pessoa. Other assistants prepared plans for a new sewage system at San Sebastian, and surveyed the 600 miles long railway from Madrid to Oviedo, in the northern district of Asturias. Various improvement schemes on the River Ebro were also undertaken. This river flows from the Mediterranean Sea, across the Iberian Peninsula, almost reaching the Bay of Biscay.

* * *

In mid-Victorian times relations between Germany and the United Kingdom were extremely cordial. The Queen's beloved Albert was of course of German origin and close family relationships existed between the two Royal Households. Reciprocal visits and exchanges of information were common place. In 1853 the Admiralty recommended JMR to their Prussian counterpart and he was commissioned to advise them on improvements to the naval dockyard at Heppens on the River Jade, near Wilhelmshaven. Within a year, in August 1854, he was summoned back to Hamburg to meet Herr Syndicus Merck, the President of the Elbe Deputation, with the result that he was commissioned to submit proposals for improvements to the Elbe navigation to accommodate 2,000 ton vessels. JMR gave the overall cost as £680,000, which included the cost of a new double dredger. Because the Elbe is frozen for two months of the year, he was also instructed to prepare designs for new docks at Cuxhaven on the mouth of the river. His estimates for this work amounted to £600,000. Although JMR paid two further visits to Germany before his death, sadly, he did not see this work completed. When a senior consultant of the present-day Rendel organisation was invited to Hamburg in the 1980's to advise the Port Authority on their construction problems, his hosts were greatly impressed when he produced JMR's original reports and plans prepared 130 years ago.

* * *

JMR's greatest concern was the proposed Suez Canal.

Probably JMR's greatest concern in the months preceding his death was the proposed Suez Canal. In 1846, the international body La Société d' Etude du Canal de Suez had been formed under the auspices of Mehemet

Ali, the Egyptian Pasha, and the French Vice Consul, Ferdinand de Lesseps. The British members of the Commission were Robert Stephenson, Charles Manley – the Commission Secretary – J.R. Mc Clean and JMR, and from its inception the engineers appeared to be at cross-purposes. Stephenson surveyed the Isthmus of Suez in 1847 and declared a preference for an extension of the Alexandria to Cairo railway across the isthmus. De Lesseps favoured the canal, but visualised an open cut like the Bosphorus, which involved dredging a channel through Lake Menzalch, 24 ft below the Mediterranean Sea, across the Isthmus of Suez to the Red Sea.

JMR sent a letter of apology saying that he was unable to attend a meeting of the Commission being held in Paris on 24th June 1856, only five months before his death, but he made it quite clear as to his and Manley's opinions.

> We propose to construct a ship canal in nearly a direct line between Suez and Port Said. The level of top water, throughout the whole length of 93 miles, to be 23 ft above low water of the Mediterranean Sea. The embankment to be carried into the deep water of the sea at each extremity, so as to avoid the necessity of dredging, and the locks to be constructed similar to the Sea Lock on the Caledonian Canal.
>
> The Canal to be supplied with water from the Nile at a point below the barrage suitable for the purpose.
>
> The execution of a Canal of this description would be so easy, that we consider the success of the project would be certain, especially as there were no contingencies greater than in ordinary engineering works of the same magnitude, and as the estimate did not exceed the amount proposed to be expended by the Engineers of the Viceroy on their original project.

A majority of the Commission elected to support de Lesseps and the open cut scheme won the day. The Canal was opened in 1869, and the intervening political manoeuvering which took place in Parliament after JMR's death need not concern us. British opposition to the Canal became political rather than scientific, but it must be said that, had JMR's scheme been adopted, many millions of pounds would have been saved in the cost of constantly having to dredge the Canal. In more recent times dredging has also been necessary in order to widen and deepen the canal. Locks would have inhibited this work, thereby hindering the progressive increase in the size of vessels able to use the canal.

A majority of the Commission elected to support de Lesseps.

* * *

The saga of the River Nene Improvement Scheme is a long and unsatisfactory story, too long to be retold in these pages. Few of the Engineers associated with the scheme over a period of many years, including JMR, escaped censure. Essentially, the difficulties seem to have arisen because the work in its entirety was not co-ordinated. Each of the several authorites involved employed their own Engineer, who individually put into effect his own ideas, without due regard for their effect elsewhere. We find Robert Stephenson writing to his principals, 'I decline to associate myself in a professional capacity with the execution of the works proposed by Mr Rendel . . . I insist that I should also formally approve and sign the plans of Mr Rendel for any works to be executed below and through the Town of Wisbech, and that such works should be completed to my satisfaction.'

It is more appropriate that we should end our review of JMR's work with mention of four bridges with which he was associated in the latter days of his life. In 1856, it became necessary to build a swing-bridge across the Nene in the centre of Wisbech. The Admiralty had insisted that the

Inverness Suspension Bridge as built by JMR.

JMR and Armstrong jointly designed an interesting hydraulically operated iron structure.

bridge should be of the opening type to allow ships to pass up the river to Peterborough, and JMR and Armstrong jointly designed an interesting hydraulically operated iron structure. By this time, JMR's third son, George, was increasingly being taken under Armstrong's wing and he was involved with this project from its inception. The bridge was 156 ft long by 40 ft wide, and the largest of its type. The bridge turned at a point about one-third along its length, where a vertical slave piston, having rollers to facilitate turning, served as a pivot. The shorter end of the bridge deck was weighted to balance the structure. A tower on the south bank housed the master piston. Prior to turning the deck, water was pumped into the system, and the two pistons seeking equilibrium, raised the bridge sufficiently to allow turning. Upon completion of the turn, the water pressure was removed, allowing the bridge deck to sink gently onto pads. Sadly, due to failing health, JMR played little direct part in the construction of this bridge which was formally opened in the year following his death. Due to lack of trade, the bridge remained closed for many years, and it was finally dismantled and replaced by a fixed bridge in 1931.

In 1850, JMR had been instructed by the Treasury to inspect the River Ness from Loch Ness to the sea, especially at Inverness where the old stone bridge had failed. Led by his old friend, Joseph Mitchell, the Highland Roads and Bridges Commissioners wanted to build a new three-arch stone bridge. Soon the two men came into conflict and Mitchell wrote

> Mr Rendel visited Inverness. I had known him when a lad with Mr Telford. He dined with me, and I furnished him with prices, and all other information. I was requested to impress upon him, in the name of the Town Council, that whatever he recommended he should not propose a suspension bridge.

In fact, initially, JMR did not propose a suspension bridge, but a wrought iron girder bridge at an estimated cost of £16,000. After much wrangling, the design was altered to a 225 ft long suspension bridge on exactly the same principle as the Montrose Bridge. JMR claimed that, not only was a suspension bridge cheaper to construct, but that it would harmonise better with the Castle and its approaches. Eventually, the work

JMR's suspension bridge across the ornamental lake in London's St. James's Park – 1856.

went ahead, but sadly, things went wrong from the very start. There were difficulties with the foundations due to imperfect cofferdams being erected, and two successive contractors went bankrupt, Armstrong lost heavily on the ironwork contract, and the costs escalated alarmingly to £30,000. The resentful Mitchell thought the whole thing something of a fiasco from the beginning, and as a result the work was not completed until April 1856. Perhaps the moral of this story is that it does not pay to ride roughshod over local opinion, when directing a scheme from far-away. One suspects that JMR was so heavily committed elsewhere that he did not have the time to supervise this project as it should have been, and Inverness did nothing to elevate his professional reputation. The hoodoo of the suspension bridge remained with him. On the positive side, the people of Inverness were delighted with the aesthetic appearance of their new bridge, and many were sorry when it was replaced by a modern pre-stressed reinforced concrete structure in 1959.

The people of Inverness were delighted with the appearance of their new bridge.

James Simpson, John Hawkshaw, John Fowler and JMR were appointed members of a Select Committee in 1856 to enquire into the state of the new Westminster Bridge. The cast and wrought iron bridge had been designed by Thomas Page to replace the 18th century masonry structure and was to be 810 ft long between abutments. It had been planned such that one side would be completed first, allowing the passage of traffic while the other side was built. The work was difficult to organise and endless delays arose, which resulted in the bankruptcy of the Contractors, C. J. Mare & Company.

JMR severely condemned the principles on which the piers were being constructed, claiming that they would be subject to decay and subsidence. He advocated a new start, in spite of £57,000 having already being expended. With recent events at Inverness in mind, he emphasied the need to employ properly constructed cofferdams and the greater use of granite in any future work. He claimed also that insufficient allowance had been made for expansion, pointing out that Telford's Menai Bridge expanded by as much as 4 ins. JMR's findings were not unanimously accepted by his colleagues and a compromise solution was eventually agreed.

JMR realised his life was drawing to a close.

By the summer of 1856, JMR realised that his life was drawing to a close. Although not ill enough to be confined, his health was failing and he increasingly showed signs of great tiredness. As if determined to fulfil his early ambitions of building a successful suspension bridge of his own design, free from the problems of Montrose and Inverness, and the disappointments of Runcorn, Saltash, Laira, Fowey and Clifton, he designed an ornamental suspension bridge to span the lake in London's St James's Park. Although, unfortunately he did not live to see its completion, it survived for 100 years. It was 140 ft long, designed to carry loads of 1100 lbs per ft run, and wide enough to carry a gun carriage.

* * *

He died peacefully on 21st November 1856 at his home, in Kensington Palace Gardens, after a short illness, surrounded by members of his family and his great friend William Armstrong. He was buried at Kensal Green cemetery.

4
The Rendel Family

Exactly one week after JMR's death Brunel took the Chair at a meeting held at the Institution of Civil Engineers. Announcing Mr Rendel's death he was generous in his praises of his former colleague, although it was generally believed that the two men were never particularly close.

'Another engineer has passed away ere his three score years and ten made their full count', he commented. 'Not dying a natural death but, we fear, alas, in a sense self-slain, undertaking the work of six men and literally dying in harness. In this case necessity had been long and far removed.'

He referred to JMR's character:

> He was a painstaking, hardworking, persevering man, desirous always to do his best and anxious always to be esteemed practical. He was a man of great energy and clear perception and correct judgement. His practical knowledge was well directed. He knew how to make good use of the scientific requirements and skill of all whose services he engaged. His evidence before committees of the House was clear and convincing, seldom failing to carry his point. Although he rose rapidly to very high position in his profession, he was as amicable and kind in his private life, as he was energetic and firm in public life.'

Then, after outlining JMR's career, he ended on a strange note,

> He was always considered a safe man – one who would seldom do anything that he had not, or others had not done before – and this is, strange to say, a recommendation with men of business, who do not understand engineering. But it is quite clear this quality is not the characteristic of a Smeaton, a Stephenson, or a Watt, or the world would make no progress.

Perhaps Brunel's words were an attempt at self-justification, for certainly JMR took none of the risks as an innovator that Brunel had taken with his broad gauge and atmospheric railways and his massive ill-fated steamships.

The family were deluged with letters from people in all walks of life – all loud in their praise for the man and his achievements and sincere in their sympathy, for few men of his generation could have boasted a wider or more varied circle of friends. The President of the Plymouth Athenaeum, of which JMR had been a long-standing member, described his passing as, 'a loss to his profession, a loss to the nation and a loss to Europe.' His staff were especially saddened, for although JMR had been severe with those he employed, always requiring strict obedience, he was far from being an unkind employer. William Armstrong appears to have

'a loss to his profession, a loss to the nation and a loss to Europe'.

taken charge of the situation at Kensington Palace Gardens, and attended to all the arrangements for the funeral, which took place on Wednesday 27th November 1856.

'Beloved Kate', as JMR always referred to his wife Catherine, was stricken with grief, but she was a woman of great strength of character and bore herself with great dignity. Since their marriage in 1828 she had personified the cliché, 'Behind every successful man is a woman'. She was very correct in her manner, ambitious for her family and warm and friendly as a hostess. She always addressed JMR as 'My darling Rendel', never using his christian name. Her father, James Harris, a native of Dartmouth, was a well respected figure in Plymouth, where he owned a house decorating and picture framing business and was a staunch member of the Congregational Church. Catherine was one of twenty-three children, which no doubt in part accounted for her strong independent character.

Between 1829 and 1843, Catherine bore ten children, two of whom died in childhood. The first born Alexander Meadows Rendel, named after JMR's architect friend, Daniel Asher Alexander, was educated at King's School, Canterbury before going up to Trinity College, Cambridge to read divinity and mathematics. Originally there was some suggestion that he should enter the church but, following his younger brother Lewis's death from consumption in 1851, he joined his father's firm and spent some time at Portland and Leith. From the moment his father died, until his own death in 1918, he guided the affairs of the business with great distinction, renaming it 'Alexander Meadows Rendel, Consulting Engineer.' Like his father he was a big man, well over 6 ft tall, but excessively modest and very 'Victorian' in his attitudes to both his family and staff.

Alexander Meadows Rendel guided the affairs of the business with great distinction.

At the age of twenty-one he fell deeply in love with Eliza Hobson, the 15-year-old daughter of Captain William Hobson, R.N., the first Governor of New Zealand. Hobson had carried out the first survey of the New Zealand coast-line and was said to have been the first white man to make friends with the Maoris. He had also carried out surveys in both the Arctic and Antarctic. One of the great treasures still retained by the Rendel family is a fine porcelain cup, said to have been part of Marie Antoinette's favourite tea set, purchased by Alexander at the 1851 Great Exhibition, and presented to his fiancée the following year prior to their marriage. 1852 was an extremely busy year in the Rendel household, for as we have already seen, JMR was at the centre of the disputed works at Birkenhead, he had been elected the President of the Institution of Civil Engineers, and the family had moved from Great George Street, to a fine house in Kensington Palace Gardens, which today forms part of the Russian Embassy.

Alexander and Leila, the name always used by the family when referring to Eliza, had nine children, two of whom later joined their father in the business. William, the second son, became a partner in 1894, but sadly he died four years later. He was succeeded almost immediately by his youngest brother, Henry Wedgwood Rendel who had already proved himself to be a brilliant engineer. He was educated at the Royal Naval College, Greenwich and trained as an engineer at Neilson's Locomotive Works in Glasgow and at Armstrong's Elswick Works. His first major responsibility was the erection of the hydraulic machinery in the Government Dockyard at Bombay. He then joined his brother George at Armstrong's Pozzuoli Works in Italy for a period of two years. In 1896 he joined his father as an Inspector and spent some time in America and on the notorious Uganda Railway, becoming a partner in the firm in 1898. At this time the Uganda Railway was regarded by the Government as of

William became a partner in 1894.

Sir Alexander and Lady Rendel in the garden of their home.

great strategic importance. A race ensued between the British and German Governments, each trying to be first with a railway to Lake Victoria. During the construction through some of the most beautiful scenery in the world, a third of the mules and a fifth of the bullocks, specially imported from India, were killed by the deadly tsetse fly, and no less than twenty-two men engaged on the project were killed by lions. Sadly, Henry died of pneumonia in 1903 at the age of thirty-seven. His death was a terrible blow to Alexander, who had always cherished the prospect of seeing the Rendel name actively perpetuated in the business.

Henry died of pneumonia in 1903.

Alexander's eldest daughter, Kate, who for many years looked after her father's accounts, married an architect, Halsey Ricardo. Their son became Sir Harry Ricardo, the founder of the world famous internal combustion engine research establishment, Ricardo Consulting Engineers Plc. Upon leaving Rugby School, Ricardo went to Cambridge to read surveying, geology and the theory of structures, intending ultimately to join the Rendel firm. He was especially popular with Alexander, because he was the only grandson showing any aptitude and interest in engineering. In his autobiography, Ricardo wrote, 'My whole soul yearned for moving machinery, for dynamic not static structures, and it was soon borne in upon me that I would never make a good civil engineer, but I cherished the hope that when I became a responsible member and partner in the Rendel firm, I would be able to extend the range of its activities to include a department of mechanical engineering which I would take under my wing'. After graduation he did join the firm, but during the First War he was permitted to pursue his great love independently, and undertook work of great national importance developing internal combustion engines for aircraft and tanks. Following his grandfather's death in 1918, he severed his connection with the firm and went his own way.

Alexander's daughter, Helen Constance, married a general practitioner, Roland Brinton M.D., with rooms at 8 Queen's Gate Terrace. Their daughter, Mary, devoted her life to politics and public service, and was elevated to the Peerage in 1966. She will perhaps be best remembered as the popular broadcaster, Baroness Mary Stocks. In later life she wrote a charming autobiography, and it is she and her cousin, Harry Ricardo, who have left us with such vivid pictures of Alexander and Eliza Rendel, and their huge circle of family and friends.

When they were first married Alexander and Eliza lived at 44 Lancaster Gate, within a few minutes walking distance of Kensington Palace Gardens. Their great friends were the Strachey family, who lived at 69 Lancaster Gate. General Sir Richard Strachey had been a distinguished Indian administrator and engineer. He was a Sapper, and for a number of years was Secretary to Lord Dalhousie, the Governor-General before becoming the Chairman of the East Indian Railway Company. To Alexander's great delight his eldest son, James, married Eleanor Strachey. James never showed the slightest interest in engineering and spent his life in the city, and doing philanthropic work as a Poor Law guardian in Kensington where he lived. In later life, he became Chairman of the Assam-Bengal Railway Company established in 1892.

James never showed the slightest interest in engineering.

Eventually, the Rendels moved to 23 Russell Square in Holborn, and acquired a splendid country residence, Rickettswood, near Charlwood in Surrey, where they spent the summer months. Eliza was not allowed to do any work whatsoever, and the children were brought up by a succession of nurses and governesses.

The family circle operated very much on a three-tier system. The top tier consisted of Alexander and Eliza, always supported by her unmarried sister Polly, who is remembered as an energetic, if somewhat eccentric lady, with a mania for buying things she did not want or could not use, such as carpenter's tools and the latest photographic equipment. The second tier was made up of the 'children', and as they married, their husbands and wives and their families. The third tier comprised the nineteen grandchildren. It would seem that, with the exception of the Wedgwood family, Alexander's brothers and sisters and their families remained very much on the fringe of Alexander's large family circle.

Alexander's sister, Catherine Emily, married Clement Wedgwood (1840-1889) of Barlaston in 1866. He was the great grandson of Josiah Wedgwood F.R.S. (1730-1795), the master potter of Etruria in Stoke-on-Trent. This was considered a 'good' marriage by the Rendel family, enhancing their social status. Clement's uncle, Josiah III married Caroline Darwin and his aunt, Emma, married Caroline Darwin's brother, the celebrated scientist, Charles Darwin F. R. S. (1809-1882). It is said that Darwin became friendly with JMR during his stay in Plymouth in 1831, the two men regularly attending meetings at the Athenaeum together. Later that year Darwin embarked upon his epic five year voyage round the world in the *Beagle,* which provided the inspiration for his controversial *The Origin of Species,* first published in 1860. JMR's friendship with Darwin undoubtedly forged the first links with the Wedgwood family, which remained close for the next seventy years.

Sir Harry Ricardo, Sir Alexander Rendel's grandson.

Clement Wedgwood and his wife Catherine Emily produced four distinguished children. Francis Hamilton became head of the great Etruria Works, Vice Chairman of the North Stafford Railway and, after the 1923 amalgamation, a director of the London Midland & Scottish Railway. Josiah Clement devoted his life to politics and was eventually elevated to the peerage. He married his cousin, Ethel Kate Bowen, daughter of Lord

Bowen, an eminent judge whose wife, Emily Frances was Alexander Rendel's sister. Ralph Wedgwood, later Sir Ralph, spent his life in railway management. After distinguished service on the Great Eastern Railway, he became General Manager of the London & North Eastern Railway in 1923. During the last war he became the supremo of British railways, as Chairman of the Railway Executive Committee. Clement's daughter, Cecily Frances married an eminent soldier, General A. W. Money.

In the years following their father's death, Alexander's three surviving brothers, George, Stuart and Hamilton, one by one joined William Armstrong, all gaining senior positions in the great industrial empire he founded with JMR's help and encouragement. In fact, for many years Armstrong treated the Rendel boys more as sons than junior partners. Armstrong was born in Newcastle-upon-Tyne in 1810, the son of a prosperous and public spirited corn merchant, who insisted that his son should be articled to a solicitor friend, Armorer Donkin, becoming a partner in the firm in 1833. Since childhood Armstrong had been interested in mechanical things. It was while observing a large iron water wheel at work during one of his frequent fishing expeditions in Dentdale, high in the Pennines, that he realised how little of the power in the falling water was being converted into useful work. From that moment he became obsessed with ideas for harnessing the latent energy in a column of water. He published his first paper in the *Mechanics Magazine* in 1838, and built a working hydraulic motor producing 5 h.p. from a 200 ft head of mains water. In 1844, he became Solicitor and Secretary to the recently established Whittle Dene Water Company, incorporated to provide Newcastle with an improved water supply. He soon became the driving force in the Company and his innovative work progressed rapidly. In 1846, he was honoured by the Royal Society, who granted him a Fellowship in recognition of his work as a research scientist. The same year he patented his first hydraulic crane, which incorporated a novel device for converting linear motion into rotary motion.

George, Stuart and Hamilton one by one joined William Armstrong.

It was at this point that JMR actively encouraged Armstrong to devote all his energies to the development and manufacture of hydraulic machinery. He visited Armstrong's father in order to reassure him that his son was doing the right thing by abandoning the legal profession, and promised that sufficient orders for hydraulic machinery would be forthcoming to keep the proposed new factory busy. On January 1st 1847, the Elswick Company was established on the north bank of the River Tyne, near Newcastle, and within a few months important orders were received from London, Liverpool and Grimsby docks.

In 1850, Armstrong was commissioned to supply five new cranes for the M.S. & L. Rly. docks at New Holland on the River Humber. The nature of the ground precluded the erection of a tower to generate the necessary water pressure, as at Grimsby, and Armstrong invented the hydraulic accumulator. This was undoubtedly one of his greatest inventions, which did more than anything else to ensure the future success of hydraulic systems. Initially, pressures of 1000 psi were generated, but these increased to 1500 psi by the turn of the century as sealing materials improved. In 1855 Armstrong and Rendel were jointly awarded a Gold Medal at the Paris Exhibition for the work they had done in this field of hydraulic engineering.

Armstrong invented the hydraulic accumulator.

During the Crimean War, the Admiralty approached JMR as an expert on the electrical detonation of explosive charges – experience he had gained during the rock blasting operations at Holyhead. They wished to blow-up Russian ships sunk in Sevastopol harbour which were causing a serious obstruction. JMR proposed that his son George and Armstrong should go out to the Crimea to supervise the work. Greatly to the annoyance of the three men the authorities rejected the proposal and the scheme proved a failure due to mishandling by the military. However, the episode had important consequences which resulted in Armstrong producing his revolutionary system of artillery. It is extremely doubtful if either man had ever given a moment's thought to the design of weapons of destruction prior to the Sevastopol incident. The story is documented in Stuart Rendel's biography, *The Rendel Papers* edited by F. E. Hamer.

> ... Armstrong was a constant visitor at my father's house whenever he came to London on business. During one of these visits I myself remember that the conversation at breakfast turned upon the exciting news that morning of the very critical battle of Balaclava, and the splendid exertions made by the sailors to bring some naval 32 - pounders into action, whose superior range determined the favourable issue of the battle.
>
> I well remember my father's outburst at the absurd ponderousness of the cannon so critically employed. My father had been the first engineer to construct, at the age of 23, a continuous iron bridge over an estuary of the sea near Plymouth. That bridge was, of course, of cast iron.
>
> Thirty years had passed since that bridge was built, and the use of cast iron for such a purpose had become utterly antiquated.
>
> My father was indignant that military engineers should have lagged so far behind civil engineering as to be still retaining cast iron for the purpose of making cannon, of which the very earliest examples, over 200 years old, had been constructed of wrought iron.
>
> He dwelt upon the apathy and backwardness exhibited by military engineers in not seeking to give to field artillery the advantages of rifling already attained in the small arms.
>
> On this eventful morning I remember my father pointing to the great lightness and strength of the small arms barrel constructed of wrought iron which rendered the use of cylindrical bullets and rifling possible.
>
> I can see my father and Lord Armstrong now before me with a bit of blotting paper between them on the table, on which Lord drew out a scheme for the enlargement to field gun size of the small arms wrought iron rifle, and I can almost hear my father's challenge to Lord Armstrong to take up the question and bring artillery up to the level of civil engineering science of the day – "You are the man to do it."

The problems facing Armstrong were both metallurgical and mechanical. After a period of intensive experimentation he became convinced that the coil system of forming wrought iron tubes by rolling bars in spiral coils and welding them together longitudinally along the edges, provided the best solution for the rifled barrel. He opted for breech loading, discarding the well established practice of muzzle loading, using non-ferrous metal for the breech. It was soon found that his guns were seven times more accurate than smooth-bore artillery and had five times the range. In 1858, the War Office adopted the Armstrong gun for field service, and the following year he received his Knighthood. Armstrong's response was typically generous, and he gave his patents connected with the gun to the government. Thereafter, the design and manufacture of ordnance was transformed, but that is outside the scope of our story.

George Wightwick Rendel (1833-1902), third son of James Meadows Rendel.

Two or three years after JMR's death, it was decided that George Rendel should leave his brother and join William Armstrong permanently in Newcastle. During the period of the works at Grimsby, he had run away from Harrow School, feeling that he had been the victim of some injustice. His father put him to work as a trainee surveyor and later moved him to Grimsby, where he became fascinated by Armstrong's hydraulic machinery. At the time of his father's death he was working at Holyhead, and the signs were that he was more interested in mechanical than civil engineering. He had a very different character to Alexander, being less academic and more practical than his brother. It appears that the two brothers were not always in agreement, and in any event George was a countryman at heart and anxious to get away from London.

Armstrong was devoted to George, who became more or less his adopted son, and placed great reliance and responsibility upon him at an early age. When the Elswick Ordnance Company was founded in 1859 he made George a partner. In 1864, George was a signatory to the deed uniting the Engine Works and Ordnance Company as Sir W.G. Armstrong & Company, a move made necessary by the decision of the Government to place no further orders for guns with Armstrong. The demand for ordnance had fallen to the point where it was no longer possible to keep both Woolwich Arsenal and Elswick supplied with work and the Government shamefully abandoned Armstrong for a period of sixteen years.

By this time George's younger brothers, Stuart and Hamilton, had joined Armstrong, together with Brunel's son Henry. Both Robert Stephenson and Brunel had died in their fifties, within two years of JMR's death, and Catherine Rendel and Armstrong did what they could to befriend Brunel's widow, Mary. She had been left in somewhat strained circumstances and Armstrong agreed to train her son at Elswick Works. Young Henry Brunel quickly matured and developed into a first class engineer, becoming a great asset on the hydraulics side of the business. The lovely story has been passed down in the Rendel family that Catherine returned from a visit to the Brunel home in Duke Street one day saying, 'she had been to see poor Mrs Brunel, who was in a terrible state, fearing that she was about to die and would be condemned to spend eternity with Isambard'! Hamilton Rendel also developed into a first class engineer. When only a child, Robert Stephenson presented him with a working model of the Stephenson Link Motion, used for reversing steam engines, 'out of pleasure in discovering that the small lad thoroughly understood the movement'. Stuart Rendel stayed in the company's London Office as the salesman negotiating contracts with the Admiralty and various foreign powers, building up an enormous worldwide connection before entering Parliament, an occupation which Armstrong considered a waste of time.

Hamilton Rendel developed into a first class engineer.

During the period when no orders were forthcoming from the British Government, the policy at Elswick was to develop the general engineering side as much as possible, seeking orders for ordnance wherever they could be found overseas, and in 1867 they became warship builders as well. During the American Civil War, the company supplied both sides, fully utilising their capacity of 50 tons of guns per week. In 1878, during a period of strained relations with Russia, the Government commandeered some Armstrong guns intended for Italy, and sent them to Malta. Shortly afterwards, breech-loading guns were again re-introduced into the British navy. Elswick had just developed a new, much improved 6 ins breech-loaded gun to George Rendel's design, which Armstrong himself considered the greatest improvement since the introduction of rifled ordnance, and the firm were given large orders by the Admiralty. Hydraulic machinery in warships became more and more complex with the improvement in guns, and the Admiralty turned more amd more to Elswick, 'for the simple reason', as Armstrong said, 'that they could hardly do otherwise'.

George Rendel went on to develop bigger and better guns, culminating in a 100 ton monster. He also played a major role in the design of warships. Initially, he believed the need was for fast ships, with the emphasis placed on fire-power and mobility, at the expense of protection in the form of armour plate. He built a series of 1,350 ton 16-knot unarmed cruisers for the Chinese and Chilean Governments. Then, in 1882, just before leaving

George Rendel went on to develop bigger and better guns.

Armstrong, he built the *Esmeralda*, a fast cruiser, with an arched steel protective deck extending from stem to stern below the level of the waterline. She had a speed of over 18 knots, a displacement of 2,974 tons and was fitted with two 10 ins and six 6 ins breech-loading guns. Intended for Chile, she was bought by the Japanese, who used her to good effect in their crushing defeat of the Russian Fleet in 1905 at the Battle of Tsushima.

Armstrong had no children of his own.

Armstrong had no children of his own, and seemed to have a propensity for recruiting able young men into his business. In the 1860's, he recruited a young gunnery officer, Captain Andrew Noble, who had been secretary to various Government committees on ordnance. He was ambitious, tenacious and a fanatical worker, often not leaving the works until midnight. Almost inevitably, in time, there was trouble between him and George Rendel. The last straw came when Armstrong proposed that Rendel should hand over the management of the ordnance works to Noble. Stuart left the company for Parliament in 1880, and two years later George left to become the first Extra Professional Civil Lord of the Admiralty, who were determined to retain as long as possible his brilliant grasp of naval ordnance and warship design. Hamilton, who disliked anything in the nature of public life, having been born with a dreadful stammer, remained quietly with the firm as an engineering designer. He never married. Perhaps his most notable achievement was to design and install the steam driven compound condensing pumping engines, hydraulic accumulators, and hydraulic machinery for operating London's Tower Bridge bascules.

Armstrong undoubtedly felt very let-down and abandoned by the Rendel brothers and in 1883, Sir W. G. Armstrong & Company became a public company enabling Armstrong, then aged 73, to retire more and more to his magnificent estate at Cragside, Northumberland. It was a late incorporation for such a large and nationally important company, and Stuart Rendel, still a major shareholder, opposed it. In fact he opposed the principle of the limited liability company altogether and wrote to Armstrong words which many feel still have some relevance today, 'I set down my belief that these new combinations of industry and this divorce of management from capital are big with consequences more momentous than even the most dreaded combinations of labour can bring forth'.

JMR's great friend Sir William Armstrong, later Lord Armstrong of Cragside.

George Rendel's first wife, Harriet Simpson, died in 1878 and in 1880 he married an Italian lady, Licinia Pinelli, whom he met in Rome during his frequent visits to Italy upon Armstrong's behalf. In 1885, after three years in London, George's health showed signs of failing and he was advised to live in a warmer climate. Eventually, greatly to the regret of his colleagues at the Admiralty, he accepted Armstrong's invitation to take charge of a new shipyard he was building at Pozzuoli, near Naples. In Italy, George and Liccinia had three children, two sons and a daughter. The youngest, George William, born in 1889, enjoyed a distinguished career in the Diplomatic Service. He served in Berlin, Athens, Rome, Lisbon and Madrid; as Minister in Bulgaria, and Ambassador to Yugoslavia and Belgium. Latterly, he was Under-Secretary of State at the Foreign Office. He was knighted during the war, receiving the K.C.M.G., and upon his retirement in 1957, he became the Chairman of the Merchant Bank, Singer and Friedlander. His fascinating life story is told in his autobiography, *The Sword and the Olive,* published by John Murray in 1957.

Stuart Rendel was quite a different character from his brothers.

Stuart Rendel was quite a different character from his brothers. He was educated at Eton College before going up to Oriel College, Oxford to read law. Although he was eventually called to the Bar, he never practised, preferring the commercial world. Shortly after his father's death he joined

Sir William Armstrong and remained in control of the London office, where he amassed a considerable fortune. In 1880, he stood as Liberal candidate for Montgomeryshire, wresting the long held seat from the Conservatives. In the House of Commons he always closely indentified himself with Welsh National causes and largely carried the Intermediate Education Act for Wales through Parliament in 1889. Later, he was elected the first President of the Welsh National Council. From the moment he entered Parliament he was popular with Gladstone, one of whose last acts before retirement in 1894, was to elevate Stuart to the House of Lords.

Lord Stuart Rendel (1834-1913), fourth son of James Meadows Rendel.

Stuart lived in a grand style and divided his time between his fine homes in London, Surrey, Brighton and Cannes. He made no secret of his acute disappointment that his wife, Ellen Hubbard, failed to produce sons. It was well known in the family that his ambition was to create a dynasty of Rendel peers. It was Stuart's attitude to his wife which so annoyed his brother Alexander, and the relationship between the two men always remained cool. Whatever Alexander did, Stuart always contrived to go one better. When Alexander purchased his 400 acre estate at Rickettswood, Stuart bought a bigger one nearby at Hatchlands. When, after the removal of the 'Red Flag Act' in 1896, Alexander purchased his first Benz car, Stuart bought a bigger and better model, and so on. His daughter Maud's eventual marriage to Gladstone's son Harry, the future Lord Gladstone of Harwarden, undoubtedly gave him enormous satisfaction.

Stuart's large shareholding in Armstrongs was almost certainly the only reason he was never given ministerial office. Upon Armstrong's death in 1900, he was the largest shareholder, and one suspects still cherished the hope that he could oust Noble and take the Chair of this great company. He made no secret of the fact that he felt there were too many Nobles taking too much out of the business. In 1903, he formally 'declared war' on Noble and made known his fear that if Noble and his friends obtained exclusive control, it was only a matter of time before the company fell into the hands of Vickers, for many years their greatest competitors. In due course that is precisely what happened.

The only member of JMR's immediate family not so far mentioned was his eighth child, Edith. She devoted her early life to philanthropic work, and became closely involved with the St Pancras Poor Law School at Leavesden in Hertfordshire. She converted a cottage on the fringe of the Rickettswood estate into a convalescent home for London slum children, and ran a holiday camp for factory girls. Dame Mary Stocks placed her at the head of her list of people that had had the greatest influence upon her own life. In due course, Edith, married Frederick Hebeler, the German Vice-Consul in London, and had five children of her own.

* * *

What an extraordinary family the Rendels were! JMR and his wife Catherine produced 10 children and they in turn produced 35 grandchildren, 56 great-grandchildren and 90-great-great grandchildren. How sad it is that none are to be found today in the great firm founded by JMR. Perhaps it is some consolation that he has one claim that neither Smeaton, nor Telford, nor Brunel, nor Stephenson can match. His name lives on in the firm, 150 years after putting up his plate in Great George Street.

What an extraordinary family the Rendels were!

The present day Directors of the firm – for they are no longer partners, but an incorporated body under the Companies Act – are continually reminded of their founder in the form of a fine 30 ins high marble bust by C. H. Mabey standing on a pedestal in the corner of their Board Room. This is in fact a copy of the original done in 1855 by the distinguished sculptor, Edward W. Wyon (1811-1885). The original stood in the Plymouth Athenaeum for many years and was thought to have been lost when the building was destroyed by Nazi bombing during the War. Recently, it has been found at Lord Stuart Rendel's former home at Hatchlands in Surrey. Prior to being handed over to the Natonal Trust in 1944, the house, with its magnificent Adam sculptured chimney pieces and elaborate plaster ceilings, was occupied by the well known architectural historian H. S. Goodhart-Rendel, who had married Stuart Rendel's grand-daughter. Wyon is also remembered for his fine portrait busts of Robert Stephenson, I. K. Brunel, Robert Napier, Henry Bessemer and Lord Dalhousie, the Viceroy of India. Members of the Civil Engineering profession are also reminded of their former President by a portrait in oils of JMR, by W. Boxall, R.A., hanging in a place of honour in the Institution's headquarters.

Genealogy of James Meadows Rendel's Family

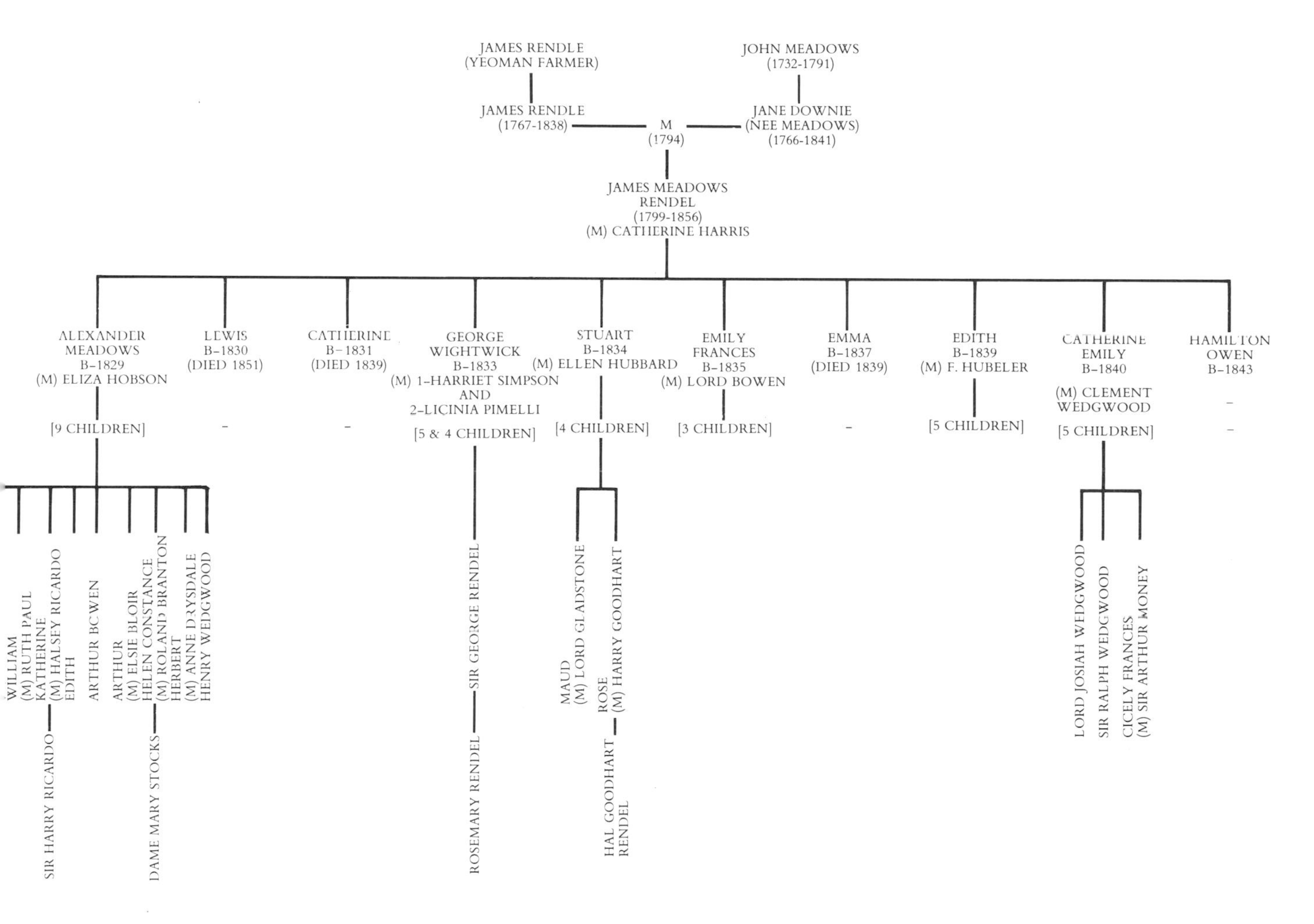

5
Sir Alexander Rendel & the Indian Railways

It was inevitable that in the midst of the Railway Mania in Britain, men of vision should dream of building railways across the great continental land masses of the world. They were the natural successors to the Elizabethan seafarers who sought to develop trade and acquire great wealth by circumnavigating the world in their little sailing ships. In Britain and northern Europe the new railways easily integrated with established centres of population and the sea ports, quickly enhancing their economies based on agriculture, new fast developing iron and coal related industries and maritime trade. In North America and on the Indian Sub-Continent and elsewhere, the railways were seen as a means of opening up the countries. Often the railway station was the first building in what is today an important, thriving centre of population. Frequently, the builders were not only pioneers, but explorers.

On the Indian Sub-Continent the railways were seen as a means of opening up the countries.

In the 1840's schemes existed on paper for building a long trans-continental railway from the northern European capitals, across the Balkans, through Turkey and Persia, to India. Concurrently, proposals were put forward for the formation of the great Indian Peninsular Railway, radiating from Bombay, towards Nagpor and Madras, and for the East Indian Railway, to link Calcutta and Delhi. Due to lack of finance and Government support, no construction work was actually started until 1852. The East India Company, the backbone of the Indian economy, was neither willing nor able to finance railways on its own, and the United Kingdom Government remained aloof from the proposals. Then quite suddenly three factors transformed the situation. Lord Dalhousi became the Governor-General and persuaded the Government at home to encourage actively greater investment in India. He advocated that some sort of Imperial guarantee be given on interest payments to entice British investors. Secondly, the Lancashire cotton industry was becoming increasingly nervous about the situation in the southern United States, as Negro emancipation became a reality. They realised the necessity of developing the Indian cotton industry as a substitute for American cotton. Thirdly, British manufacturers generally were facing growing competition in Europe, and saw India as a new outlet for their goods.

The vast Indian Peninsular and East Indian Railways were given a 5 per cent guarantee on their first lines, and naturally this was demanded by all promoters who followed. The 1857 Mutiny, whose suppression was greatly assisted by the new railways, proved to be a turning point. It not only established the need for, but removed all grounds for argument against the extension of the railway system. With the abolition of the East India Company's rule, the Indian Government authorised eight new companies and 5,000 miles of line.

JMR referred to Indian Railways at some length in his Presidential Address.

JMR referred to Indian Railways at some length in his Presidential Address given before the Institution of Civil Engineers on 13th January 1852. His first duty in the Chair, however, had been to confirm the election of his friend Thomas Brassey as a member of the Institution. The two men had become friendly six years earlier whilst building the Birkenhead and Chester Junction Railway and by the time Brassey was formally recognised by the profession, he had completed successfully over fifty railway contracts in the United Kingdom, France, Spain and Norway.

Rendel commented,

> In India two lines of railway have been commenced with every prospect of success, being under the patronage of the Indian Government, the works conducted by English Engineers, and executed by English contractors, most of whom are connected with this Institution.
>
> Our Associate, Mr R. Macdonald Stephenson, the first projector of these lines, went to Calcutta in 1843 for the purpose of directing the attention of the Indian authorities to the value of railway communication, in such a country and climate. After encountering the labour and vexation, always attendant on what is commonly called 'innovation', he succeeded in establishing the practicability of a line from Calcutta to Delhi, a distance of upwards of eight hundred miles, and in showing that such a line, whilst it promised to open up new, as well as to increase the present commercial resources of the country, would also stimulate the natives to fresh exertions, and reduce the machinery of government. Above all, that it would hasten the accomplishment of the great mission, which our rule over the millions of our fellow-creatures, inhabiting that vast region, has imposed upon us, of raising them from an almost animal, to an intellectual condition.
>
> Mr R. Macdonald Stephenson is now Resident Manager for the East Indian Railway Company, in the construction of a line of one hundred and twenty miles, connecting Raneegunge with Calcutta, forming the first section of the great railway. Associated with him is Mr George Turnbull, as the principal resident Engineer under my direction as Engineer-in-Chief.

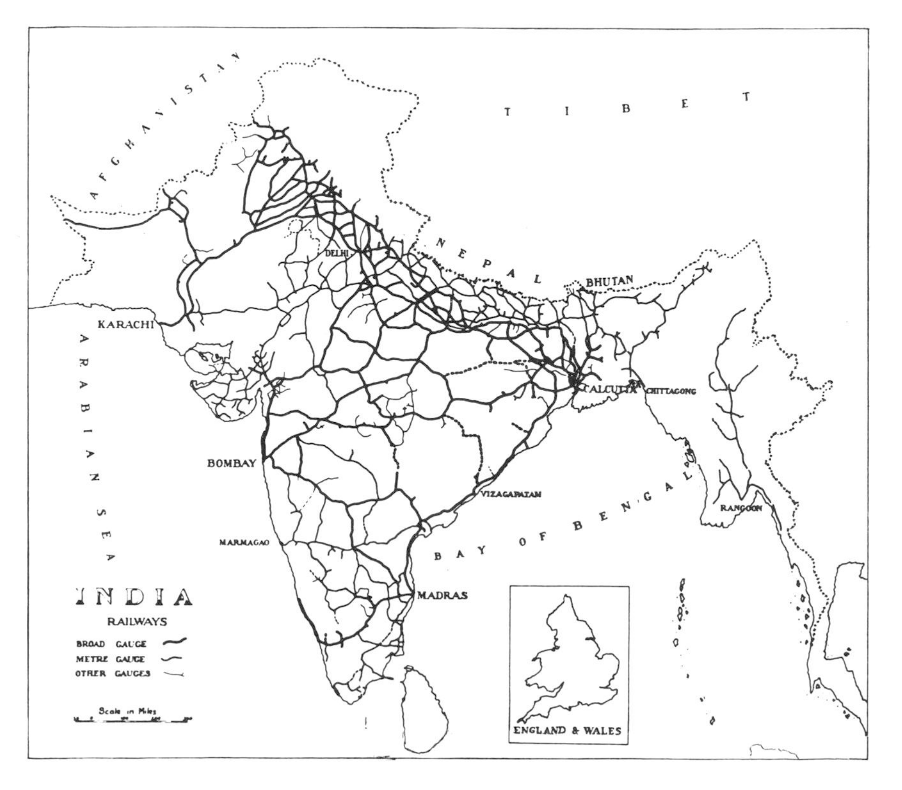

Development of the Indian railway system in the sixty years 1856-1916.

The Keul Bridge, East Indian Railways, having 9 spans each 150 feet long.

We do not know how JMR came to be appointed Engineer-in-Chief of this great project, although he had had tenuous connections with India for a number of years. He never actually visited the country, but sent out his assistant, Charles Greaves, to manage his affairs. We can only speculate that the answer lies in JMR's professional and personal relationship with Robert Stephenson. Shortly after JMR's Presidential Address, James Berkley, the Resident Engineer of the Great Indian Peninsular Railway was entertained to a luncheon in London attended by leading members of the profession. In his speech Berkley referred to the fact that the Indian Government and the two new railway companies were 'able to refer to Mr Robert Stephenson in all things at all times', adding that, 'Assistants sent out to India were all expressly chosen by him.'

The biggest engineering problems facing the promoters of the new railways were undoubtedly the wide rivers to be crossed, which changed dramatically in character with the seasons, and the very unsatisfactory nature of the soil. Greaves stayed on in India for five years, and in a report home he described the whole of the Bengal plain as, 'nothing but a sea of mud, there was hardly a stone as big as a coco-nut, or a hill as high as a house. It is a wonder, having regard to the softness and looseness of the soil, that Calcutta remained where it was.' Being minutely aware of JMR's career, and bearing in mind especially his achievements at Laira and Grimsby, which were not dissimilar to the Indian problems, it seems perfectly natural that Stephenson should have recommended his friend to the Indian Government.

Greaves described the Bengal plain as 'nothing but a sea of mud'.

Indeed it was the very nature of the Hooghly River, on which Calcutta stands, that originally brought JMR into contact with the Indian authorities. Even before he moved his office to London in 1838, he had been approached by an organisation calling itself the Steam Ferry Bridge Association of Calcutta. They wanted to build a 'Floating Bridge' across the river, and after prolonged negotiations a contract was placed with Acraman & Morgan's Bristol Iron Works for two iron vessels. Greaves went to Bristol to supervise their construction, and in January 1842 he accompanied them on their long voyage round the Cape. Unfortunately a dispute had arisen between the ferry company and the Bengal Marine Board, and work on the slipways and approach roads had to be abandoned. All the materials delivered to Calcutta, including the two 'Floating Bridges' were subsequently sold at public auction. During his sojourn in India, Greaves lived on the banks of the Hooghly observing and preparing various schemes, which included carrying out a survey for the proposed

Great Western Railway of Bengal. In 1854, before any work was started, the company was absorbed by the East Indian Railway and the line was eventually built under Rendel's supervision.

JMR was also Consulting Engineer to the Government of Ceylon Railways.

JMR was also Consulting Engineer to the Government of Ceylon (Sri Lanka) Railways until his death. The construction of a 74½ miles long line from Colombo to Kandy was originally projected in 1847, but the work was not commenced until 1863 and opened to traffic in 1867. John Coode was responsible for the original survey, but upon JMR's death Sir Charles Hutton Gregory was appointed Consulting Engineer. We do not know whether JMR's appointment in Ceylon preceded that in India or not, but it is noteworthy that both systems adopted the 5 ft 6 ins gauge. In view of the Stephensons' long drawn-out 'Battle of the Gauges' with Brunel, one naturally wonders what Robert Stephenson's reaction was to the decision, and whether the responsibility for its selection rested wholly with JMR. It must not be forgotten that the only other major overseas railways he was responsible for surveying were the 600 miles long Madrid to Oviedo Railway in Spain and the Recife Railway in Brazil, both of which were broad gauge. One report suggests that it was Lord Dalhousie himself who insisted upon the 5 ft 6 ins gauge, but it is unlikely he would have taken this step without prior consultation with the Engineer-in-Chief. Nevertheless, had both men been able to foresee subsequent events in India, they would almost certainly have adopted the standard 4 ft 8½ ins gauge, avoiding the complications of the metre gauge introduced on many secondary lines.

In September 1847, representatives of the East Indian Railway Company Board in London, accompanied by a team of engineers carefully selected by JMR set sail for India. It had already been agreed that the East India Company should grant the railway all the land it required on a 99 years lease, free of all costs, and that the railway company, its officers and servants were to be subject to the control and supervision of the East India Company at all times. Because of the extremely difficult crossing of the Hooghly river, the line was to start from Howrah, opposite Calcutta, the first stage being a single track line to Pundooah and the Collieries at Raneegunge, a distance of 120 miles.

The line to Pundooah was opened in August 1854 and at the railway company's Annual General Meeting held in London, the Chairman, Mr H. A. Aglionby M.P., reported 'he could not talk too highly of the indefatigable exertions and untiring energy of Mr Rendel; the advanced state of the works would speak for themselves. Not only was the line from Calcutta to Raneegunge, a distance of 121 miles, almost finished but an extended advance was actively going on to Rajmahal, a further increase of 120 miles.' JMR in his reply commented,

> There are engineering difficulties to contend with in India, which people at home cannot possibly conceive. Yet I am bound to say that the works executed are equal to any of the kind done in this country; several large bridges have been built over nullahs and rivers near Hooghly, and on exceedingly treacherous, sinking and shifting ground. Yet no failures have happened nor have any accidents taken place, although since the planning of the railway heavier floods have risen in Bengal than have been witnessed since the days of Clive. Before the end of the year the works will bring their rails to the Raneegunge coal fields and great profits will acrue when this is completed. On the opening of the line, the engines and rolling stock will be found equal to anything of the sort in England. The Directors of the East Indian Company had readily met and concurred in all my suggestions and by the extraordinary exertions of the engineers, a survey to Allahabad had been completed in six months. Within four years the line will be advanced to this populous and important town and seven years hence your railway will be running to Delhi.

JMR's great Indian triumph. The Sone Bridge across the River Son near Patna (Bihar), as built in 1855, with 28 spans each 150 feet long.

Undoubtedly, the greatest of JMR's achievements in India was the Sone Bridge across the River Son, near Patna in Bihar, 200 miles east of Allahabad. It was 4,536 ft long and had 28 wrought iron spans, each 150 ft long. In this work, as well as the great bridges at Keul and Hullohur, he co-operated with the distinguished architect Sir Matthew Digby Wyatt (1820-1877). Upon JMR's death, Wyatt was officially appointed architect to the East Indian Railway. Due to damage sustained during the Mutiny in 1857, parts of the Sone Bridge had to be rebuilt, and single line traffic only operated until 1870.

* * *

The period following JMR's death was a period of great anxiety for the family.

The period immediately following JMR's untimely death in November 1856 was a period of great anxiety for the Rendel family. It was agreed that JMR's sons Alexander and George should try and continue the business, but because of their age and limited experience the Admiralty appointed new consulting engineers to complete the great works at Holyhead, Portland and in South Africa. The brothers did, however, receive an interesting letter from Samuel Smiles, the Secretary of the South Eastern Railway, in January 1857, seeking their advice concerning a proposed new pier and low water landing place at Folkestone for cross-Channel ferries. They reacted quickly and presented their written report on 18th February 1857. They proposed an open timber viaduct from the railway station to the low water-line and a 1,100 ft long breakwater, with a spur to check the drift of shingle. The plan appears to have been opposed by the company's own engineers and was shelved. Samuel Smiles was of course best known as a distinguished writer of engineering biography. Subsequently, he sought permission to write JMR's biography, but the invitation was politely declined.

One of William Armstrong's first acts after his friend's death was to write to Robert Stephenson, seeking his assistance to secure the future position of the two young men and thereby ensure the continuation of the Rendel business. Within a few months, the appointment of Alexander Rendel as Consulting Engineer to the East Indian Railway was confirmed, thus establishing an association which was to last for sixty years. Right up until his death in 1918, Alexander was personally and directly involved in great engineering works in India. Even at the time of his death, in his ninetieth year, he was busy designing a spectacular new bridge over the Lower Sone.

Alexander decided it was necessary that he personally visited the country.

In 1857, India was for a time paralysed by the Indian Mutiny, and work on the new railways was brought virtually to a standstill. Alexander decided that it was necessary that he personally visited the country to become acquainted with local conditions and in particular to examine the problem of crossing the River Hooghly between Calcutta and the railway terminus at Howrah. He inspected the works along their entire length to Cawnpore, beyond which it was not then safe to proceed. He introduced iron girders in the construction of bridges, replacing the brick arches formerly used on smaller bridges, and greatly facilitated the movement of supplies and materials up country by the introduction of shallow draught steamers for crossing rivers. Some of these were built in Calcutta, others were shipped out from England. In 1858, work was resumed on the railway, and considerable progress was made on the line between Allahabad and Cawnpore.

JMR had written to the 'Calcutta Englishman' as early as 1855, suggesting that a giant tubular bridge on the plan of Stephenson's Menai Bridge be constructed across the Hooghly River at Calcutta. He offered to send his son out immediately to deal with the preliminaries. But when he suggested that the cost might exceed £½ million, the project, like Dickens's immortal fat boy 'went to sleep again'. The idea of using steam ferry boats had also been discounted because of the peculiarity of tides, where eddies and flurries abounded. At certain times the current in the middle of the river ran in the opposite direction to that along the banks.

Alexander proposed a 'Floating Bridge' – not a vessel warping its way along an iron chain, as designed by his father, but a huge pontoon bridge. The roadway, 30 ft wide and 2,000 ft long, was supported on twenty-six flat bottomed iron pontoons, each 80 ft long by 26 ft wide. They were placed 50 ft apart and moored by heavy chains to a cable carried across the river from bank to bank. On each side of the river two of the pontoons were moveable, giving a 155 ft wide opening for the passage of ships. At each end of the roadway was a 110 ft long platform, one end resting on the shore, the other on a pontoon. By this means, the 22 ft rise and fall of the tide could be accommodated. The cost of the bridge, which was built by Sir Bradford Leslie, was estimated at £125,000. The construction of a new subsidiary port on the River Mutlah, near Calcutta, was also approved.

Whilst Alexander was in India, approval had also been obtained for a 200 miles long branch line to Jubbulpore, with the object of ultimately providing an end on connection with the Great Indian Peninsular Railway's line from Bombay. An extension of the main line from Delhi to Lahore was also sanctioned, but later this scheme was handed over to the newly established Punjab Railway.

All this fierce activity was not without its dangers. In 1858, the Chief Engineer of the Allahabad to Jubbulpore extension, Mr Evans, and his two assistants, Limnell and Campbell, were attacked by a posse of natives while having their mid-day meal. Colin Campbell managed to escape on horseback and, by hard riding, chased for many miles by the rebels, reached the safety of Allahabad. The next day he led a party over the route of his escape, the official report recording, 'how he stuck on his horse over such country is one of the marvels of horsemanship'. The rebel leader, a mutineer named Runmust Singh, had poor Evans head cut off and he made Limnell carry it until he became exhausted. Runmust Singh then ordered his men to cut off Limnell's head, but this they refused to do, saying we have already killed one sahib, you must kill the other, which he did by shooting him down. Subsequently the country was scoured by Gurkhas and Runmust Singh was captured and hanged at Rewah.

Throughout 1859 construction work proceeded apace.

Throughout 1859, construction work proceeded apace on several

Kalka Station, North West of Delhi showing the E. I. R. broad gauge train alongside the Kalka to Simla railway's narrow gauge train.

sections of the railway simultaneously. The chief difficulty lay in the transport of materials up country. A terrible cholera epidemic ravaged the Rajmahal district and for some weeks eight to ten per cent of the coolies employed died weekly. Before the disease disappeared more than 4000 labourers had succumbed. The engineer's reports confirm what an appalling time they had, but, by the end of 1859, considerable progress had been made. The 24 miles between the River Adjai and Sainthea station and the remaining section in the Beerbhoom district had been opened to traffic and the section to Rajmahal was almost completed. From Rajmahal to Colgong works were in an advanced state and good progress was made in sections as far as Monghyr, with work having been started on the Jamalpur tunnel. The only bar to more rapid progress was the need for bridges and permanent way materials, which could not be advanced sufficiently quickly. In the North West Provinces also, work was already going on as far as Agra, and arrangements were in hand for getting possession of the land necessary for the entrance into Delhi. By the end of 1859 some 360 miles of line were in service, 1,172,852 passengers and 299,424 tons of freight were carried during the year and the year and the company operated 19 passenger and 30 goods engines. Two years later the length of line in operation had been increased to 603 miles.

Throughout 1863 further sections of the main line were completed and on 1st August 1864 a service was inaugurated up to the banks of the river Jumna at Delhi, a distance of 954 miles from Calcutta. Within the space of two years the company's revenue doubled to £1.12m. The last link in this great railway was the Jumna Bridge at Allahabad which took 8 years to construct.

Alexander proposed the construction of the Chord line.

During 1864, Alexander proposed the construction of the 252 miles long Chord line from Raneegunge to Luckeeserai, thereby shortening the distance between Calcutta and Delhi by 147 miles and avoiding the necessity of doubling the existing single track main line along the right bank of the River Ganges which in due course became known as the Loop Line. It also gave improved access to the large coalfields in West Bengal, benefiting not only the Company, but the public and the shipowners at Calcutta, counterbalancing the advantages claimed for Bombay as a steamship port. Thomas Brassey and George Wythes, the greatest contractors of the day, were awarded the contract, but work progressed extremely slowly and the line was not completed until 1871. This was in part due to a serious trade depression in 1867 which affected the whole Indian economy, and serious flooding which rendered parts of the railway impassable.

The Board in London deemed it necessary to carry out a reassessment of its position and asked Alexander to visit the railway as soon as possible and to report on the situation. He spent four months in India, returning home in 1868. He found that capital expenditure in the previous two years had been running ahead of what he termed, 'the needs of the natural development of traffic', and he advocated cut-backs in approved expenditure amounting to £¾ million. The question of building a railway bridge across the Hooghly river to give direct access to the centre of Calcutta was shelved, as was the doubling of the line from Gahmar to Allahabad. He criticised some aspects of the operation of the railway, pointing out that the mileage run by the rolling stock was out of proportion to the revenue earning work done. He recommended new statistical methods designed to control the costs of working the line, and these soon proved to be of enormous benefit to the Company. A Chief Superintendent of the railway wrote of Alexander Rendel's work in later years, 'if asked to point out the most important service rendered by him, I would mention the railway statistics initiated by him; they became a most valuable guide to the proper conduct of railways and the chief basis for economies in working'. Alexander had proved himself not only a brilliant engineer, but a wise and prudent economist, policy maker and business manager. He was a good 'all-rounder' in what had become a huge organisation. He was held in such high regard by the Indian Government that in 1872 he was appointed as Consulting Engineer to the Secretary of State for India.

Alexander was a good 'all-rounder' in what had become a huge organisation.

In 1876, Edward, the Prince of Wales, visited India and travelled extensively in a specially built Royal train, which was used as the Vice Regal train for the next 25 years. His busy itinerary was published in the *Gazette of India,* dated 22nd April 1876.

> On the East Indian Railway. His Royal Highness the Prince of Wales travelled in January 1876 from Howrah to Bankipore, and thence to Benares, also from Cawnpore to Delhi; and from Delhi to Ghaziabad, on going to the Punjab, and from Ghazibad to Agra in returning thence. In February, His Royal Highness travelled from Agra to Aligarh, and in March from Cawnpore to Allahabad and on to Jubbulpore.

Concurrently, Alexander also visited India. His prime concern was the re-opened question of a bridge across the Hooghly river between Howrah and Calcutta, and the conversion of the Howrah property to deal with the growing wheat, seed and coal traffic. During his visit he took the opportunity of travelling over the whole railway system, whether in the Royal train or not, we do not know, but many important questions were settled locally with the railway's officers.

In 1878, Richard Strachey, now a Lieut. General and a member of the Council of the Secretary of State for India, conducted joint talks with the Government and the Railway authorities to discuss the acquisition of the Railway by the Government. Alexander Rendel was requested to accompany his friend to India because, quite apart from the question of purchase, it was becoming increasingly urgent to ensure that future capital requirements were correctly identified to meet the anticipated development of traffic. The outcome of the talks was a Bill which received the Royal assent in August 1879, transferring ownership of the Company to the Indian Government. The total length of the line at this time had grown to 1504 miles, an average rate of construction of 60 miles a year during its 25 years existence.

The first problem Alexander was called upon to resolve under the new regime was the provision of adequate means of transporting goods from the railway at Howrah to Calcutta and vice-versa. The warehouses and docks were mostly on the Calcutta side if the Hooghly River, and goods arriving at Howrah by rail could only reach them by using the 'Floating Bridge'. A connection to Calcutta by rail was therefore greatly to be desired. Alexander was convinced that a crossing of the river should be established 24 miles north of Calcutta, giving the East Indian Railway a direct link, via the Eastern Bengal Railway, with the Calcutta Port Trust jetties. Government approval was quickly obtained, and the splendid bridge, probably the climax of Alexander's bridge building career, was formally opened by the Viceroy on 21st February 1887. The ceremony coincided with the Queen's Golden Jubilee, and the bridge was aptly named the Jubilee Bridge. It measured 1200 ft between abutments and carried two lines of railway track. The central double cantilever was 260 ft long, and the two main side spans were each 420 ft long. The height of the bridge was 53 ft above the datum on the two river piers. Both the builder, Bradford Leslie and Alexander Rendel were appointed Knight Commanders of the Indian Empire. Shortly afterwards another magnificent Rendel bridge was opened across the Ganges at Benares.

A connection to Calcutta by rail was greatly to be desired.

In 1889, Robert Wigram Crawford, the highly successful Chairman of the Company for 35 years, died. He was replaced by General Sir Richard Strachey and he and Sir Alexander Rendel arrived in India together in February 1890. Probably the most pressing problem awaiting them upon their arrival was the long outstanding question of the proposed 281 miles long Grand Chord Line – the name given to the most direct route

Attock Bridge built in 1885 across the River Indus. The railway on top deck and the road beneath.

between Calcutta and Delhi, connecting Bengal with United Provinces, Central India, Bombay and the Punjab. Originally surveyed in 1850, the Indian Government of the day preferred the more circuitous Loop Line along the densely populated banks of the river Ganges. Crawford reopened the question in 1886, claiming that the Grand Chord would not only give relief to the growing traffic on the existing main lines and shorten the direct route from Calcutta to Delhi by 50 miles, but would

The Sukkur Channel Bridge across the River Indus.

e Sone Bridge, the longest India, was opened in 00.

consolidate the great railway system in the Ganges Valley and give improved access to the important coalfields at Palamow. Strachey put forward new proposals to the Indian Government in 1890, but sanction to proceed was not given until 1895.

The first section of the line built was from Gya to Moghalsarai, a distance of 125 miles, and included a spectacular bridge over the River Sone at Dehri, and the branch line to Palamow. The Sone Bridge, the longest in India, and at the time the second longest in the world, was opened in February 1900. Its overall length over abutments was 10,054 ft and comprised 93 spans, each of 100 ft long. Through services on the Grand Chord between Calcutta and Delhi were established in 1906.

The last major project for the East Indian Railways in which Alexander was personally and directly involved was the building of the new locomotive works at Jamalpur on the Ganges. A site was chosen near the town of Monghyr, which for long had been known as the 'Birmingham of the East' and where a plentiful supply of skilled labour was available. Today, Jamalpur might be termed the 'Crewe of India'. A site of 100 acres was rapidly developed and by 1906 there were over 20 acres of covered workshops, foundries and rolling mills, complete with their own electricity generating station. In its own way, Jamalpur was just as much a great engineering achievement as Alexander's spectacular river bridges. At this time he was over 75 years of age, he had been Consulting Engineer to the East Indian Railways for over 45 years, and he was getting tired. The long sea voyages to India were getting too much for him, and increasingly he handed over his Indian work to assistants.

By no means was Alexander's work on the Indian sub-continent confined to the needs of the East Indian Railway. He had been totally integrated with that company and fulfilled admirably the role of Engineer-in-Chief, and in many respects even that of Deputy Chairman. Other Indian Railways used his services as a Consultant, dealing with specific engineering problems as they arose. In 1876, he designed the Alexander Bridge across the Chenab River for the North Western

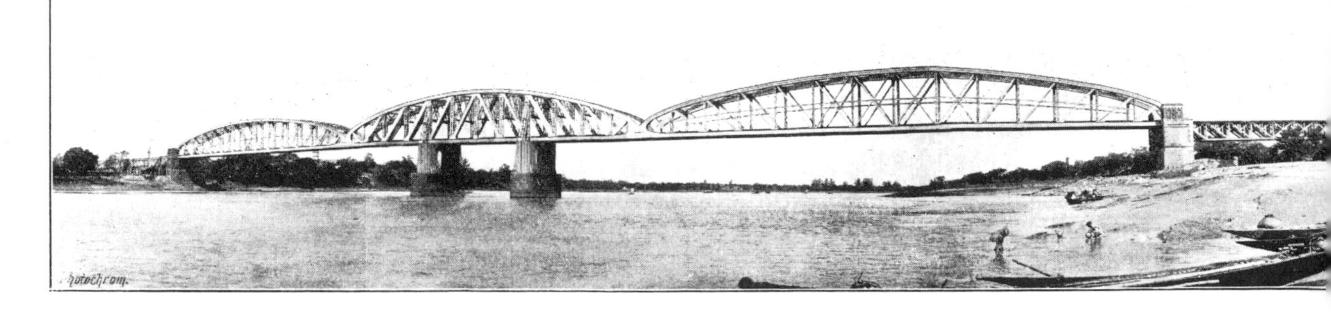

The Jubilee Bridge built in 1887 carrying the East Indian Railway over the Hooghly River.

Railway. As built this was 9,088 ft long and comprised 64 spans each 133 ft 6 ins in length. In the 1880's he undertook surveys for the 5 ft 6 ins gauge Bengal and Nagpur Railway, which in a sense was sandwiched between the East Indian Railways and the Great Indian Peninsular Railway with its termini at Seetarampore, 138 miles from Calcutta, and at Nagpur and having a branch line to Jubbulpore. In 1885, he designed the Attock Bridge across the River Indus for the North Western Railway. This had 2 spans 302 ft long and 3 spans 250 ft long, the railway running on the top deck and a roadway on the lower deck. This was followed by the Sukkur Channel Bridge across the Indus having spans of 290 ft, 230 ft and 88 ft. In 1889, he built the beautiful Lansdowne Bridge across the Indus. Until overtaken by Sir John Fowler's Forth Bridge, this was the longest cantilever bridge in the world, having a cantilever span of 790 ft between abutments. He designed also a number of bridges for the metre gauge Assam Bengal Railway. In Burma he carried out extensive surveys on the Rangoon to Mandalay Railway, and in 1901 supervised the construction of the spectacular curved trestle viaduct at Gokteik on the Lashio branch. The viaduct was 2260 ft long and at its highest point was 320 ft above the bottom of the valley. The detailed design of the steelwork was subcontracted in this instance to the Pennsylvania Steel Company.

In 1857, when Alexander was appointed to succeed his father in India, 500 miles of railway had been already authorised. At the time of Sir Frederick Palmer, his successor's death, in 1934, the State owned railways operated 17,000 miles of railway, the privately owned independent lines 14,000 miles, and Burma and Ceylon (Sri Lanka) added a further 2,000 miles. Of this total, the Rendel partnership had acted as consulting engineers for a staggering total of 31,000 miles, surely a record which cannot be equalled anywhere in the world. In the same period they designed and supervised the construction of over 200 major bridges on the Indian sub-continent, having some 3,000 spans.

... the Rendel partnership acted as consulting engineers for a staggering total of 31,000 miles of railway.

Nor was Alexander's work in India solely with civil engineering and bridge building projects. Even with the rapid development of railways, the great Indian rivers continued to dominate the social and economic scene, and for many years there were those who thought the Government would have been well advised to spend more money on the development of the waterways, and less on the railway. In his early years of office, Alexander constructed several 'Floating Bridges' of an improved design, the most famous being that across the Hooghly River at Calcutta. In 1857, he designed and built a large railway ferry across the Ganges for the North Western Railway. The landing stages were 100 ft long by 40 ft wide and were connected to the shore by pontoon bridges able to rise and fall with the river. The steam powered ferry boat had three lines of railway track fore and aft converging on a turn-table in the centre of the vessel, the total length of track being of sufficient length to accommodate the average train. In the period from 1886 to Alexander's virtual retirement from the

Placing the girders of the Barrakur Bridge in position on the Grand Chord Railway.

Indian scene he designed and supervised the construction of 15 paddle wheel ferry boats for the East Indian, the Bengal and North Western, the Bengal Dooars and the Bengal and Nagpur railways.

Concurrently, a Marine Department was established at the London office and a variety of special craft were designed for overseas clients. In 1891, a grab dredger was supplied to the Madras P.W.D. A similar 800 ton vessel went to Rangoon, and another 900 ton suction dredger went to Madras. In 1905, a large bucket dredger was also built for Madras. Various special purpose tugs and harbour craft, and shallow draught lake steamers were designed for the Calcutta, Madras and Rangoon Port Authorities and for the Uganda Railways for service on Lake Victoria.

Khushalgarh Bridge across the River Indus built in 1908. Railway on top deck, roadway beneath.

Landsdowne Bridge across the River Indus built in 1889. At one time the largest cantilever bridge in the world with span between abutments of 800 feet.

But we must leave India and return to London, where in the midst of his heavy commitments to the Indian Railways, Alexander carried out some of his most distinguished work in London's dockland, which placed him at the very top of his profession. Unlike his father, he took little or no part in the proceedings of the Institution of Civil Engineers, and rarely sought the company of his peers. He was a lone wolf in his profession keeping his business affairs to himself and enjoying an affluent, but very private life with his large family and few close friends. He worked single handed as the Principal of his business, and it was not until his 70th year that he took an outside partner. His business and private life-style clearly became increasingly affected by that of the Indian Raj.

The history of London's enclosed docks is long and involved. Thomas Telford was the first great engineer to be called upon to give evidence before a Parliamentary Select Committee, but the credit for building the first part of what became the largest dock complex in the world went to William Jessop. He was responsible for the West India Docks started in 1799. John Rennie built the London Docks and the East India Docks, and designed the new London Bridge, which was built by his sons. In 1824, the St. Katherine's Dock Company was formed on the site of the old St. Katherine's Hospital near the Tower of London. The Company received its Act in 1825, and Telford was appointed Engineer. Fifty years later, what was to become the largest totally enclosed dock system in the world was authorised, and the promoters, the merged London and St. Katherine's Dock companies, appointed Sir Alexander Rendel their Engineer. He was well known to them, having been a Consulting Engineer to the London Dock Company since 1857. Originally, the intention was to cut a ship canal from the existing Victoria Dock at Blackwall to Galleon's Reach, in order to give an entrance three and a half miles lower down the river, and to save the navigation of two rather difficult bends in the river. Alexander pointed out that new and larger docks and locks would inevitably be required in the future, and that large sums of money could be saved if they were constructed at the same time as

. . . the merged London and St. Katherine's Dock companies appointed Sir Alexander Rendel their Engineer.

Shershah Bridge across the River Chenab. Built in 1890 with seventeen spans each 200 feet long.

the canal. The scheme was quickly revised and the depth of water in the docks was increased from 26 ft to 30 ft. The width of the passage between the old and the new dock was also increased to 80 ft. The whole system as proposed by Alexander formed a magnificent dock 1¾ miles long covering an area of 175 acres, and having 87 acres of enclosed water. Two entrances were provided, the original into the Victoria Docks at Blackwall Point and a new one at Galleon's Reach, below Woolwich. The latter entrance was in the widest part of the river and was protected by two guiding jetties leading in to the entrance lock. The lock was 800 ft long by 80 ft wide and was fitted with three pairs of wrought iron gates, capable of admitting not only the largest sea-going merchant ships afloat, but any of the Royal Navy's Ironclads.

Immediately beyond the lock was an entrance basin of about 9 acres, leading into the main dock by a passage 300 ft by 80 ft in which was placed a pair of gates similar to those at the lock, enabling the basin to be converted into a 9 acre dock. The passage was spanned by a swing bridge carrying the East Ham road across the dock. The main dock had an area of about 75 acres and was about 1¼ miles long and of a uniform width of 490 ft between the copings.

The dock walls throughout the works were constructed entirely of Portland cement concrete made and deposited on site. The gravel for the concrete was excavated on the site and this was one of the first civil engineering contracts carried out entirely by mechanical means, extensive use being made of the recently introduced Ruston Steam Navvy excavators. It was also noteworthy as the first major contract to be lit entirely by electricity. The dock walls were 40 ft high by 5 ft thick at the top and 19 ft thick at the base, requiring a half million cubic yards of concrete, and incorporating 80,000 tons of Portland cement.

... this was one of the first civil engineering contracts carried out entirely by mechanical means.

On the south side of the main dock two large dry docks were constructed, the smaller one 420 ft by 68 ft and the larger 510 ft by 84 ft wide. The sills of both were 22 ft below Trinity high water. Under the river entrance to the old dock a 1,800 ft long tunnel carried the North Woolwich branch of the Great Eastern Railway. A further double track of railway crossed the passage from the old to the new dock on the largest swing bridge in the United Kingdom. It had a span of 90 ft and weighed over 860 tons.

THE ROYAL ALBERT DOCK.
MR. E. M. RENDEL, M.I.C.E., ENGINEER; MESSRS. LUCAS AND AIRD, CONTRACTORS.
(For description see page 454.)

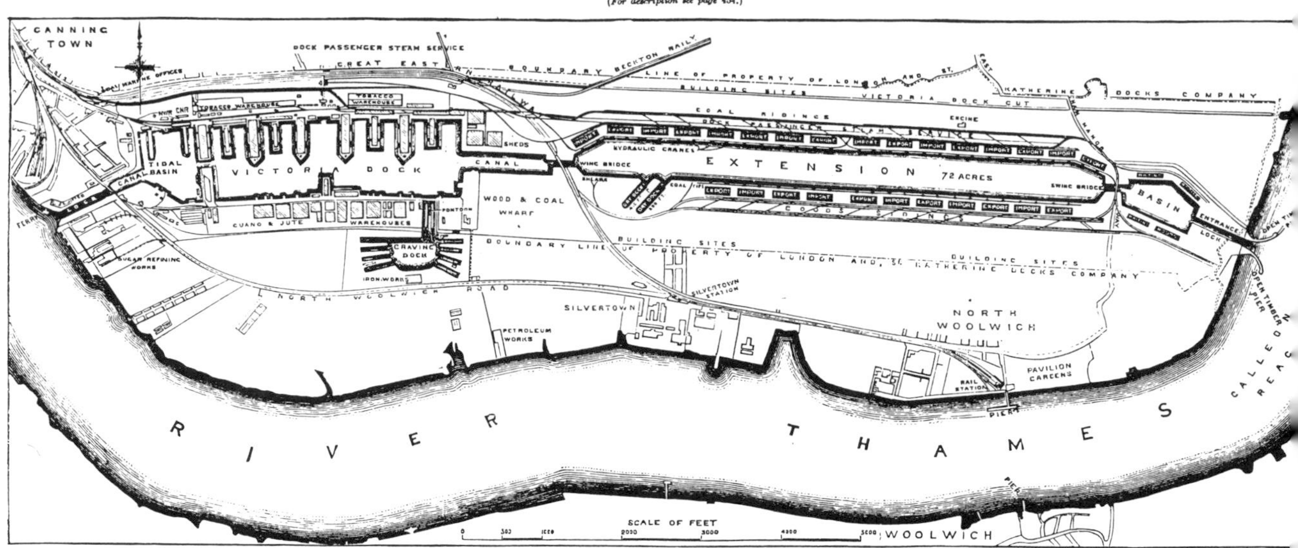

The dock was formally opened by the Duke and Duchess of Connaught representing the Queen on June 24th 1880. Twelve steamers carrying 3,000 invited guests sailed down the river from the Embankment and lined both sides of the new Albert Dock. The Royal couple then entered in the Trinity steam yacht *Vestal* to the accompaniment of a Royal Salute. After a brief inspection of the North Quay the whole company enjoyed a sumptuous luncheon. How very proud Alexander must have felt that day.

Sir Alexander Rendel's Royal Albert Dock opened on 24th June 1880, showing its connection with the River Thames and the existing Victoria Dock.

The Duke and Duchess of Connaught entering the Royal Albert Dock in the Trinity House steam yacht 'Vestal' during the opening ceremony on 24th June 1880.

6

The Palmer and Tritton Partnership 1918-1934

The majority of people remember the year, if not the actual dates, which subsequently proved to be turning points in their lives. Undoubtedly, the year 1883 would have been regarded by Sir Frederick Palmer, K.C.M.E., C.I.E. (1862-1934), a Past President of the Institution of Civil Engineers, as a turning point in his life. In 1883, he applied successfully for the post of Assistant Engineer on the East Indian Railway at the then very adequate salary of £45 per month. The final decision for the appointment had rested with Sir Alexander Rendel, the company's consulting engineer, and during the next 30 years, great mutual respect and friendship developed between the two men.

Frederick Palmer was born at Llandovery, in the southern foothills of the Cambrian Mountains, in January 1862. He was the third son of George Palmer, who at the time described himself as a carpenter, although later he established a small contracting firm working mainly for the Great Western Railway in the South Wales district. George Palmer and his wife, Sarah, are believed to have come originally from North Lincolnshire. There is even a suggestion in the family that, as a young man, he was employed on JMR's great harbour works at Grimsby, erecting the huge timber cofferdams.

In 1877, at the age of fifteen, young Frederick Palmer was articled to John Lean, the Neath District Engineer of the Great Western Railway for a period of three years. The premium payable amounted to £150, a not inconsiderable sum of money in those days for a struggling artisan to invest in his third son's training. During his articles, Frederick is known to have been employed on the Taff Vale extension of the Newport, Abergavenny and Hereford Railway and to have acquired considerable skill in lettering and colouring his drawings. But he was an ambitious young man and anxious to broaden his horizons, and he soon began to look beyond the Welsh valleys for suitable employment. Following his successful interview at the East Indian Railway Company's offices in Nicholas Lane, he secured passage in the S.S. *Australia* and sailed from Gravesend in November 1883, spending his first Christmas away from the family at sea.

We know little of Frederick's early years in India, working on the Ghat Section of the line, but the indications are that the young engineer found life agreeable and full of promise. In 1896, he was appointed Personal Assistant to the Chief Engineer, Mr Frederick Ewart Robertson (1847-1912) based in Calcutta. Roberston went out to India in 1868 and had been closely associated with Alexander Rendel since his appointment as engineer responsible for the train ferry across the River Indus at Sukkur. In 1889, this was replaced by Rendel's beautiful Lansdowne Bridge and Robertson was credited with the novel design of the equipment used for the erection of the huge 820 ft span cantilever. He was rewarded by being made a Companion of the Order of the Indian Empire, and in a subsequent paper before the Institution of Civil engineers, he described the design and construction of the bridge in detail, which earned him a coveted Telford Medal. Robertson held the office of Chief Engineer of the East Indian Railways for eight years, before accepting the Presidency of the Egyptian Railway Board. In 1898, he returned home to the United Kingdom and joined Sir Alexander Rendel with whom he remained a partner until his death in 1912.

Robertson remained a partner until his death in 1912.

In 1895, Frederick Palmer learned that the Grand Chord line had been sanctioned by the Indian Government and that a start on the first 200 miles long section from Serai to Gya was imminent. By this time Robertson had left for Egypt and he approached his successor, E. H. Stone, requesting that he be considered for the job as Resident Engineer. The Chief Engineer is reputed to have replied, 'Yes, I don't see why you shouldn't have it; you won't make a bigger bloody fool of yourself than the others!'. He was put in charge of the section of the line which included the spectacular two miles long, 93 spans bridge at Dehre-on-Sone referred to in the last chapter. This was exactly the type of challenge that Palmer had been seeking. The 91 twin-well foundations of the bridge had to be excavated in the dry season, leaving the main stream of the river to flow through a narrow central channel. The wells were built up 20 ft or 30 ft above the river bed level and then sunk down through the sand and silt by grabbing out from the inside. Theoretically, a simple straightforward system of building bridge foundations, but one requiring great skill in execution in the soil conditions found in India.

During this work, Palmer made his headquarters at Allahabad. It was here that he became great friends with George Huddleston, the biographer of the railway, who subsequently became the Superintendent of the line. In later life, both men recalled their happy recollections of this period, during which most week-ends were spent in the hills shooting leopards and fishing. Not only did Palmer firmly establish his reputation as a civil engineer, whilst in Allahabad, but as a leader of men. Furthermore, he demonstrated his appreciation of the importance of economics in the planning and execution of major projects of this nature. The criterion he used for estimating the annual income of a new railway line, although simple, was widely used by all Indian railways for many years. He laid down that the gross annual income could be safely taken as one Rupee per year per head of the population living within a 5 miles wide strip on either side of the new alignment.

Palmer's task on the Grand Chord was completed in 1900.

His task on the Grand Chord was completed in February 1900 when the great bridge was formally opened. He was then given extended leave and for only the second time in sixteen years returned to the United Kingdom. During the voyage home in the S.S. *Arabia*, Palmer met a wealthy American, John Elliott Mason, and his very attractive daughter, Florence Elizabeth, who were enjoying a world tour together. It appears that the couple fell very much in love during the voyage, and a few months later a further meeting was arranged in Paris. Marriage was obviously discussed, but the 38-year old British engineer did not exactly satisfy Mason's aspirations for his daughter and initially he opposed the marriage. Florence was not only a very beautiful young lady, but she was intelligent and had considerable artistic talent. She had recently obtained a degree in philosophy at the University of California in San Francisco, and one suspects that Mason had in mind nothing less than a 'blue-blooded' English aristocrat for his talented daughter. But Frederick Palmer was as determined in affairs of the heart as he was in his professional life, and the marriage eventually took place in Calcutta in 1903.

Towards the end of 1900, during his long leave, which he spent mostly with his brothers in Wales, he decided that it was time he considered the next stage of his career. In contemplating marriage, he obviously wished to find employment that would enable him to establish a permanent home in a congenial locality and a job which would entail the minimum amount of separation from his new wife. Two suitable appointments were vacant at the time: firstly, the Chairmanship of the Rangoon Port Trust in Burma; and secondly, that of Chief Engineer of the Port of Calcutta. Sir Alexander Rendel had been asked by both authorities to name suitable candidates, and he unhesitatingly made strong recommendations in Palmer's favour. Frederick wrote to his fiancée, 'I may tell you that Sir Alexander has been a great friend to me for many years and somehow or other seems to have got a very exaggerated view of my ability and energy.'

Fortunately for Palmer, fortunately for Rendel, and fortunately for the profession, he did not secure the primarily administrative job in Rangoon, but was appointed Chief Engineer to the Port of Calcutta, a post he held for eight years. After a period of settling down, during which time his marriage to Florence was solemnized in St. Paul's Cathedral in Calcutta on 23rd December 1903, he began the enormous task of planning the reorganisation and enlargement of the privately owned port to meet future projected needs. In order to understand and appreciate the magnitude of this undertaking we cannot do better than to quote from Palmer's Presidential Address, given before the Institution of Civil Engineers in London in 1926.

> The city and port of Calcutta are situated on the left bank of the River Hooghly, some 80 miles from the sea. The site of the original settlement was chosen as being the place nearest to the sea with a long reach of deep water reasonably immune from cyclones, some of great intensity which frequently occur in the Bay of Bengal. In 1848 a Government Committee recommended the construction of docks at Kidderpore, with provision for vessels 'of an average burthen on 400 tons', but up to the time of the creation of the Calcutta Port Trust in 1870, the only berths for vessels consisted of four small tee-headed jetties on the riverside, equipped with cranes and sheds. These jetties were connected up and extended from time to time until the whole available frontage of 4,400 ft was utilized. Construction of an enclosed dock was commenced in 1884 at Kidderpore, about three miles below the jetties. The works comprised an entrance lock, 400 ft long by 60 ft wide, entering into a tidal basin, from which there was separate access to the river through an 80 ft entrance with a single pair of gates.
>
> No. 1 Dock, 2,750 ft long and 600 ft wide, with twelve berths was opened in 1892, and thereafter Dock No. 2 has been proceeded with as required. The entrance lock has also been lengthened, by means of inner and outer caissons, to 510 ft. The accommodation since added within the docks consists of six berths for general goods and ten berths for coal exports. There are sheds with a floor-area over 1 million sq. ft for the storage of grain, and two large warehouses of four storeys in which 4.5 million lbs of tea can be stored.
>
> In 1906 it became apparent that further extensions of the dock accommodation would be required in the near future, and a scheme was prepared under which a site, some 1,750 acres in area and with a deep water frontage to the river of over 3,000 ft was then acquired for future developments. The scheme included two entrances from the river, a dock 2,400 ft long and 1,000 ft wide parallel to the river and two branch docks, 5,750 ft long by 800 ft wide and 5,900 ft long by 600 ft wide, one at either end of the main dock. The total water-area of the ultimate development was to be nearly 200 acres.

> The taking up of land for dock and railway works in India is a compulsory acquisition by the Government, the compensation to owners being based on the value at the date on which the land is notified in the Gazette as being required for public purposes. It is therefore a matter of some importance, especially in the neighbourhood of large cities, to maintain the utmost secrecy in the preparation of proposals involving land purchase, in order to avoid exploitation in values. The preparation of the scheme was consequently carried out in my own house, and the gazetting of nearly 3 sq. miles around the southern and eastern sides of the city came as a great surprise to Calcutta.
>
> Beyond the acquisition of the land, nothing was done for some years.

In all probability, this last sentence provides the answer as to why in 1909 Palmer sought pastures new. Undoubtedly, since his marriage, he had found life in India extremely pleasant. He and Florence enjoyed a lovely home at 101 Garden Reach. They were waited on hand and foot by faithful Indian servants, and lived a full and splendid social life. No doors in fashionable Calcutta were closed to them. Frederick was also able to indulge his favourite pastime of dressage and owned several fine horses. But, he was still the same ambitious man who went out to India in 1883. He remained as determined as ever to get to the very top of his profession. He was not content to just manage a large port. He was convinced that his 1906 plans provided the solution to the problems of the Port of Calcutta, and became frustrated when Government sanction was withheld. In later years Frederick Palmer's foresight in designating land for future development became apparent. The Halada dock in India, 55 miles downstream from Calcutta, was not proceeded with until after the Second World War. Similarly, land at Tilbury, reserved in the same way, became essential when the Container Terminal was built after the war.

Palmer was not content to just manage a great port.

In 1908, the Government at home passed the Port of London Authority Consolidation Act, which effectively combined all Thames dock companies into a single authority. Shipowners and users had been complaining about the state of affairs in the Port of London for over twenty years, pointing out the urgent need for energetic action to improve the navigable channel of the Thames between Gravesend and the Pool of London, as well as the docks themselves. The channel depth was insufficient to allow vessels to come to anchor without the danger of running aground, which meant they had to keep moving even in London's then notorious fogs. The shipowners demanded that the channel should be at least 26 ft deep at low water, the same depth as the Suez Canal. Attempts had been made as early as 1887 to alleviate the situation, when the East and West India Dock Company built deep water docks down the river at Tilbury, but they had been warned by a group of eminent engineers that the tidal basin as designed would simply become a mud trap. Indeed, the first Orient Liner to use the dock, the S.S. *Garonne,* stuck fast in the mud, and the new experimental submarine, H.M.S. *Nautilus* was very nearly lost during her trials, caught in the mud trap. But the problems of London's docks were not only of a technical nature, but also financial. Several of the old dock companies were virtually bankrupt, and it was undoubtedly during this period that the seeds were sown, which resulted in a long and complex history of poor industrial relations. Ultimately, these proved to be a deciding factor in the decline of the docks in the post-Second World War years.

The new Port of London Authority advertised the post of Chief Engineer in June 1909, and Frederick Palmer took the third decisive step in his career, applying successfully for the appointment. Fortunately, he was in England on leave at the time, and events moved quickly. He was offered the job on July 8th and four days later reported for duty. He settled his American wife, Florence, and their small son John, then five years old,

Frederick and Florence Palmer with their son John, photographed in India – circa 1907.

in a new home in Blackheath, and energetically set about the long term re-organisation of the docks and associated structures. In less than eighteen months from joining the P.L.A., on 31st December 1910, he submitted to Lord Devenport, then chairman of the P.L.A., his now famous report entitled, 'Proposals for Improvement and Extension of Dock Accommodation in the Port of London'.

The report was acclaimed as one of the most far-sighted produced in the field of civil engineering.

The report, which proposed an expenditure of £14.5 million (at least £100 million at present day values) was acclaimed as one of the most far-sighted produced in the field of civil engineering. It was undoubtedly a climax in Frederick Palmer's career, for it is one thing to design a major new scheme on a 'green field' site, but he demonstrated his great skill and ability by proposing ways and means of carrying out a whole series of improvements on a huge complex and largely obsolete existing site. Within twenty years, his scheme made London the premier port in the world. He not only demonstrated the necessary skill and vision, but he had a sufficiently strong and persuasive character to convince Lord Devonport, his successor, Lord Ritchie, and then the government, that his schemes were viable and worth carrying out.

* * *

The story of the construction of the King George V Dock, the enlargement of the Albert Dock, the improvements to existing installations in the Pool of London and the building of a new dock and passenger landing stage at Tilbury must wait whilst we catch up with events in Sir Alexander Rendel's partnership. Mention has already been made of the tragic deaths of Sir Alexander's two sons, William and Harry, and of Frederick Ewart Robertson, Palmer's former chief in India, being made a partner. In 1898, following William's death, Robertson and Harry Rendel were made partners, Robertson being treated as the senior by virtue of his age and experience, and the name of the firm was changed to Sir Alexander M. Rendel & Company. After Harry's death in 1903, the style of the firm was again changed to Rendel & Robertson.

Robertson was undoubtedly a very talented engineer. Not only had he a profound knowledge of engineering science and practice, but, as is often the case in the lives of clever men, he had many other accomplishments. He was a fine organ-builder and several churches in India were indebted to him for their organs. In 1897, he published a two volume treatise on the subject. He was a talented painter and spoke fluent French and Arabic, compiling an Arabic vocabulary whilst in Egypt. In many ways he was the very antithesis of Alexander, who cared little for music or the arts. Unlike Alexander, Robertson took an active part in the affairs of the Institution of Civil Engineers, being elected a Member of the Council in 1911. But he remained a 'Boffin' and lacked any real stature. He was quite unable to match the relationships Alexander had established by his wisdom, integrity and personality with the senior ranks of civil servants, especially those at the Colonial Office and India Office, who relied upon him absolutely. Lady Mary Stocks described her childhood recollection of Robertson as, 'a nice old gentleman who took us to the pantomime every Christmas'.

Robertson brought some business to the firm, mainly from Egypt and the Sudan, and, of course, he remained *persona grata* in India. He was involved in the construction of the new Red Sea Hills railway from Atbara to Port Sudan which was opened in 1906, shortening the route to the coast by 1,000 miles. In 1910 he designed a road and rail bridge over the Blue Nile linking Khartoum North with Khartoum, followed by the Kosti Bridge across the White Nile on the 400 miles long Khartoum to El Obeid railway. But according to Harry Ricardo, his grandfather was not very happy in his partnership with Robertson, and relied more and more on Seymour Biscoe Tritton, a mechanical engineer who had been with the firm as chief of staff for some years. Alexander was approaching eighty and becoming deaf, and although his mind remained as alert as ever, he left more and more of the executive work to Tritton. Perhaps this is an indication of the depth of Alexander's despair, because Ricardo tells us that both his grandfather and Robertson took a very poor view of mechanical engineers in general. In their eyes the civil engineer was the artist, the other, a mere mechanic belonging to a much lower position in the hierarchy. A short time after Robertson's death in November 1912, at the age of 65, Tritton was made a partner, and the name of the firm was again changed to Rendel and Tritton.

He was born in 1860, the son of a regular army officer, and was educated at Haileybury College and University College, London. He received his technical training as a premium apprentice at R & W Hawthorn's works in Newcastle-upon-Tyne, and then spent some time at sea as a marine engineer. In 1885, he went out to India as assistant locomotive superintendent on the Bengal and North Western Railway and was placed in charge of the company's new workshops. Later, he joined the Indian State Railways as works manager of the former East Bengal Railway workshops at Kanchrapara, which employed over 10,000 men. In due course he was promoted to the post of district locomotive superintendent of the Northern Bengal State Railway responsible for the workshops, running a department and their flotilla of river craft.

He fitted into Sir Alexander's team admirably, continuing the traditions of the British Raj in the London office. He was the instigator of the firm's increased activity in the fields of mechanical and marine engineering and built up a considerable reputation as a locomotive engineer, later becoming President of the Institution of Locomotive Engineers. He was an acknowledged expert upon standardisation and was responsible for developing the very important engineering inspection side of the business. His work designing ever bigger and more powerful dredgers, tugs, and other special marine craft was recognised by his membership of the Institution of Naval Architects.

Tritton built up a considerable reputation as a locomotive engineer.

Learning of Robertson's death and Tritton's appointment to the partnership, Frederick Palmer was shrewd enough to guess that Sir

Alexander, then in his eighty-third year, was probably apprehensive and concerned about the future of the business. Although he continued his daily routine of making the five mile journey from Ricketts wood to Horley Station by carriage and pair, and then briskly walking from Victoria Station to his Westminster office, he realised that he could not go on for ever. He must also have asked himself the question, would the Indian Government and other important clients accept a mechanical engineer as heir apparent? Although none of the correspondence of this period appears to have survived, Lady Palmer recalled in later years that her husband had written to Sir Alexander in effect saying, 'I wish there was something I could do to help you'.

In 1913 Palmer joined the Rendel partnership as Sir Alexander's deputy.

The outcome was that, in April 1913, Palmer resigned from the P.L.A. and joined the Rendel partnership as Sir Alexander's deputy and heir apparent, the name of the firm being changed yet again to Rendel, Palmer and Tritton. He was retained by the P.L.A. as a consultant on a fee basis for a period of five years, so that he could act in case any contractor's claims had to be settled. He secured the handsome guaranteed minimum fee of £4,000 p.a. from the partnership. Sir Alexander is reputed to have commented, 'I never thought I could catch such a big fish!' Palmer's ambitions were such that it seems doubtful in retrospect whether he would have been content to stay for long with the P.L.A. He was by now a man of enormous stature in his profession, and it seems unlikely that he would have been prepared to remain an employee indefinitely. If Sir Alexander had not made the offer, then somebody else would have come forward. It must remain to Tritton's great credit that he accepted the position of third man, when he had expected to be second. Although a very different character from both Sir Alexander and Palmer, it was he who ensured that the three men worked together in a spirit of co-operation and harmony for the good of the firm. His reward came later, when he became Senior Partner upon Frederick Palmer's death in 1934.

The Palmers established a very comfortable life-style, but suddenly, in August 1914, peace was shattered by the largely unexpected outbreak of the First World War. They occupied a fine early Victorian house named Morden Hill House, standing in 3 acres of ground in Blackheath, and it was during this period that Frederick became very interested in gardening. He employed a full time gardener living in a cottage in the grounds and a chauffeur. The domestic staff in the house comprised a cook, a kitchen maid, a parlour maid and a house maid. Florence was a marvellous hostess, her dinner parties became legendary and the house was constantly full of guests.

The majority of Frederick's schemes in the Port of London were frustrated by wartime restrictions, and although he remained the only P.L.A. official to have access to the Board Members Dining Room, he was largely ignored by his successor, an omission which hurt him very much. However, he was no respecter of persons nor of government red-tape and the leadership he quickly established in the partnership did much to restore its declining prestige in the eyes of Whitehall. In 1916, he accepted Lloyd George's invitation to become an Honorary Adviser to the Ministry of Munitions. In due course, he became Chairman of the Ministry's main Finance Committee and enjoyed direct access to the Minister upon request. The following year, Mr Winston Churchill took over this vitally important Department of State, and in his now well-known style set about making changes which upset Frederick. He felt that the authority and 'kudos' which he had previously enjoyed was at stake and he remained determined to retain his honorary status.

He received the following handwritten conciliatory letter from the great man.

Dear Mr Palmer,

After very careful consideration of the arrangements which are appropriate for the Finance Group, I am convinved that it would be of great advantage if you would continue in your position as Chairman of the Munitions Works Board. I hope you may find it possible to undertake this work and I am sure your doing so would conduce to the public interest.

You are well aware of the importance of the work of this branch in the past, and I am informed that the new arrangements will add considerably to the scope and consequence of the duties attached to it.

I trust therefore that I may count on a continuance of your assistance and I may add that this wish is strongly shared by many of your friends here.
Yours sincerely
Winston S. Churchill.

Sir Seymour Biscoe Tritton K.B.E., Senior partner, Rendel Palmer & Tritton, 1934-1937.

Inspite of appeals from his friends and colleagues he decided to resign his appointment as Director-General of Contracts Finance, Ministry of Munitions. Florence's opinion of the incident was that Churchill had decided against having top men in the Ministry in an Honorary capacity, for 'How can you control a man who works for no salary?' she asked. But Frederick soon realised that he had made a serious mistake, one which he never repeated in the remaining seventeen years of his life. Thereafter, he devoted a great deal of his time and money befriending and entertaining people in high places. This incident almost certainly delayed his knighthood.

During the War the firm were appointed consultants to the War Office and Ministry of Munitions for railway work. Seymour Tritton looked after this work and was responsible for the design and inspection of rolling stock and railway materials employed on all fronts. He supervised the procurement and supply of 9,000 miles of track, 3,400 locomotives, many of which were requisitioned from the main line railway companies, and 72,000 wagons. No doubt somewhat to Palmer's chagrin, he was knighted for his War Work, receiving the K. B. E. in 1918.

Sir Alexander Rendel died in 1918 aged 89.

Sir Alexander Rendel died in January 1918 aged 89, from a cold which turned to pneumonia. His grand-daughter, Lady Stocks commented, 'Fortunately for him there was no penicillin in those days and no resuscitation. He swallowed his usual cascara pill, went to sleep and did not wake again.' His beloved wife Leila had died a few years earlier, and despite his enormous family, many of whom had become scattered by War Service, his last years were spent in great loneliness, looked after by Anne Drysdale, his son Herbert's widow, to whom he became very attached. Rickettswood and 23 Russell Square were sold, after his death, which resulted in the disintegration of what had become known as the 'Rendel Connection', and things were never again quite what they had been, either in the firm, or in the family. Again quoting Lady Stocks's autobiography, she referred to her generation of the family as having 'shared the experience of having lived in a golden age, singularly free from personal tensions and sex problems, and wholly at variance with the modern tendency to decry the family as an obsolete social group'.

Throughout the last thirty years of his life, Alexander remained a slightly eccentric English gentleman, steeped in the traditions of the British Raj. He had few friends outside the family and there were few callers at his home. Right up until the war, he never went out in his car, without being accompanied by a horse and carriage, more often than not leaving the chauffeur, or his grandson, Harry Ricardo, to get the car home as best they could.

Ernest Pool, a draughtsman in the office at the time of Alexander's death – later to become the senior engineer in the Steel Bridge Department – recalled his last image of the great engineer.

> Sir Alexander would occasionally totter into the drawing office looking like a venerable old country parson in his wide round brimmed low crown black hat. He had a draughtsman working on the design of the Lower Sone Bridge for the East Indian Railways. His scheme for brick arches was not favoured by the younger members of the firm and while Sir Alexander's scheme was being developed, others were working on the approved steelwork design. Sir Alexander was thus kept happy, involved with a major engineering project to the moment of his death.

The death of Alexander Rendel in January 1918, and the ending of the Great War in November 1918, marked the beginning of a new era for Rendel, Palmer & Tritton, exactly 80 years after their founder, James Meadows Rendel put up his plate in Westminster. Frederick Palmer became the senior partner of the firm as planned, and Seymour Tritton assumed the number two rôle.

* * *

The partners had every reason to look forward to a period of great prosperity.

The politicians were proclaiming, 'A land fit for heroes' and there were still, as yet, no signs of the sun setting on the British Empire. The world map was still dominated by areas shaded red. France and Germany had been devastated both economically and politically by the war, Russia was in the throes of bloody revolution, and the mighty American dollar, although increasingly flexing its muscles was still a comparative fledgling. The R. P. T. partners had every reason to look forward to a period of great prosperity.

One of the first major post war decisions taken by the partners was to invite Sir Robert Richard Gales, another member of the old Indian railways school, to become a partner in the firm. He was born at Littlehampton in Sussex in October 1864 and was educated privately. He did his technical training at the Royal Indian Engineering College at Coopers Hill, and this was followed by a period of practical training working under Sir John Fowler on the Forth Bridge. He went out to India in 1886, and after a spell with the Indian Public Works Department, he held posts as Assistant Manager with both the North Western Railway and the East Coast Railway. In 1903, he was appointed Engineer-in-Chief on the construction of the Curzon Bridge at Allababad, and three years later became Engineer-in-Chief of the Coomoor-Oomcamund Railway on the southern tip of India. He then superintended the construction of Sir Alexander Rendel's magnificent 15 span 5170 ft long Hardinge Bridge across the Lower Ganges at Sara. His subsequent paper describing the bridge given before the Institution of Civil Engineers earned him an Indian Premium and a Telford Gold Medal. During his eighteen years with RPT he was responsible for the Uganda Railway Extension, the precast concrete box girder Falluja Road Bridge over the Euphrates in Iraq and the spectacular Lower Zambesi Bridge.

In 1920, Frederick Palmer sold his Blackheath residence and purchased Crowhurst Place, a magnificent moated and timber-framed Tudor house near Lingfield in Surrey. The house had been the home of the Gainsford family for many generations, and was said to have been the over-night stopping place for King Henry VIII when travelling to Hever Castle to visit his beloved Anne Boleyn. F.P. acquired the property from Consuela, the Duchess of Marlborough, who had modernised the house and greatly improved and extended the gardens during her short period of occupation. The Palmers fell completely in love with the place, and for the next fourteen years it became not only their cherished home, but a centre of hospitality. The rôle of Squire in such lovely surroundings was very much to F.P's taste. The week-end house parties arranged to coincide with the Lingfield Race Meetings were often enlivened by a mixture of

Sir Frederick & Lady Palmer's home, Crowhurst Place, near Lingfield, the birthplace of many important Rendel connections.

Florence's American friends and members of F.P.'s Welsh family. These proved to be especially happy, long remembered occasions. Clients and potential clients from all over the world were lavishly entertained at Crowhurst, enabling him to build up many informal and close relationships, which proved to be of inestimable value in subsequent negotiations. Frequent visitors included Lord Devonport, the Chairman of the P.L.A; the Rt. Hon. Jimmy Thomas, Secretary of State for the Colonies; Sir James Carmichael, the Crown Agent; Sir Clement Hindley, Chairman of the Indian Railways Board; and Sir Herbert Walker, then General Manager of the London & South Western Railway, shortly to become one of the main constituents of the Southern Railway.

Whilst in India, F.P. had become an inveterate 'Club Member', and in the post-war years he became a member of three London Clubs, the St Stephens, the Oriental and the Carlton, his sponsor for the latter being his great friend Sir Robert McAlpine, the building contractor. He regularly attended the many social functions associated with his professional work, and although not a particularly good public speaker, he could always command his audience's attention. His choice of anecdotes was always a delight to his listeners. In every way, he was the very antithesis of his predecessor and mentor, Sir Alexander Rendel. He took an active part in the Proceedings of the Institution of Civil Engineers. Traditionally, at that time, the only speakers at meetings were those in high positions, partners in consultants, senior Local Government officials, but rarely contractors. This custom prevailed until just before the last war. Proceedings were published, but not for many months after the session. Ordinary members in the provinces and overseas frequently did not receive their copies for twelve months. This matter became of serious concern and the Council needed much prodding before the system of Journals was introduced.

The London & South Western Railway wanted to modernise and extend Southampton Docks.

One of the first major commissions given to the firm in the short-lived post-war boom was received from the London & South Western Railway, who wanted to modernise and extend Southampton Docks. At this time Southampton was rapidly replacing Liverpool and Plymouth as the main Trans-Atlantic Passenger Terminal. The company proposed building a series of ten finger piers to provide twenty 1,000 ft long berths for large ocean liners. F.P. successfully persuaded them that it would be more advantageous to substitute the finger piers with one 7,500 ft long straight quay wall. Later, as the work progressed this was increased to 17,000 ft. He recommended single storey transit sheds, which would not have been possible in the finger pier layout, due to the restricted space between the fingers. The scheme was completed in 1926 at a cost of £4.7m. F.P. always maintained that his scheme saved the company £1.2m.

Concurrently with this work F.P. took a personal interest in a project to establish an electric street tramway system in the Southern Italian town of Taranto, famous for its large modern Naval Base. Effectively, his proposals amounted to an economic and feasibility study in which he advised the promoters, the Taranto Tramway & Electrical Company, to accept the conditions which the local authority wished to impose upon its operations. More than sixty years had elapsed since James Rendel undertook his great works at Spezia and forty years had passed since his son George re-equipped the Italian Navy with fast modern ships, but the name Rendel remained revered and respected in Italy.

The huge new London dock which F.P. proposed in his plans presented to the P.L.A. in 1910, was finally completed using direct labour in the summer of 1921. Although the dock, comprising 9,000 ft of quays and an 850 ft long lock, was started almost immediately in 1911, the outbreak of war delayed its completion. The King sailed into the new dock in the Royal Yacht to formally name it the King George V Dock, in a ceremony not dissimilar from that performed forty years earlier when the adjacent Albert Dock was opened by the Duke of Connaught. F.P. had the honour of being presented to the King as the Designer. Four years later, when Kirkpatrick, his successor as Chief Engineer to the P.L.A., retired, the authority brought him back as its Consulting Engineer.

In 1923 RPT received its first job from an oil company.

In 1923, Rendel Palmer & Tritton (RPT), received its first job from an oil company. The then Anglo-Persian Oil Company commissioned the firm to examine the report on the reasons for the shoaling of the channel across the bar of the Shatt-Al-Arab at the head of the Persian Gulf, and to make recommendations for a new channel to improve navigation. F.P. went out to Basrah personally and decided that the main cause of the silting was the width of the channel below Fao, which had been increased to 600 ft. His subsequent report dealt also with the amount of silt brought down the great tributaries of the Shatt-Al-Arab, the Tigris and Euphrates flowing through Iraq and the Karum through Iran. He recommended a new dredged channel following the long curved alignment of the natural flow and having a depth adequate for vessels drawing up to 26 ft. He specified that the bottom width of the channel was not to exceed 300 ft with 1 in 40 flat side slopes, and proposed that two drag type suction dredgers be acquired to undertake the work and maintain the channel. The dredgers were to be capable of excavating to a depth of 45 ft and operated during ebb tides in order to be able to lift the silt from the bed and discharge into the stream. He took the precaution of recommending that the dredgers had large capacity hoppers in case the ebb disposal proved unsatisfactory.

The Port of Basrah Director General, Colonel John Ward, was infuriated by RPT's intervention and insisted that the estimated £450,000 necessary to complete the work should be provided by the British Government, in order that the work could be undertaken departmentally by the Basrah Port Trust. Soon after the project got underway, the Port Trust got into financial difficulties, and somebody suggested to Ward that a shorter and cheaper connection to the deepwater might be possible. Against the advice of the then Chief Engineer to the Port Authority, Ward mistakenly authorised the dredging of a new channel at an angle of approximately twenty-five degrees to that recommended by

A post war aerial view of the Royal Albert and King George V Docks, London.

F.P. From that moment, the whole scheme was doomed to failure and was finally abandoned in 1943. Fortunately, neither the British Government, nor the Anglo-Persian Oil Company, both of whom suffered financial loss, held RPT in any way responsible for the fiasco.

In 1920 F.P. visited China as the British representative on the Shanghai Harbour Board. Following the submission of a preliminary report published in 1921, the Government additionally appointed him as Engineer Member of the Yangste Kiang River Commission, to report on proposals for the improvement of navigation on the river. It had been suggested in an earlier report, that the huge river, which meanders through the heartland of China for over 2,000 miles, should be dredged to give free access to ocean going vessels of 20 to 25 ft draught up river as far as Hankow, some 600 miles from the sea. It was further proposed that, beyond that point, the river should be dredged to enable the draught of river craft to be increased from 10 to 15 ft. Although F.P. enjoyed enormously his second 1923 visit to China, when the revenue cruiser *Pin Ching* was put at his disposal, his subsequent report effectively advised the River Commission not to waste their money. In his opinion, the scheme was neither econonically nor commercially viable.

F.P. effectively advised the River Commission not to waste their money.

The second Government inspired assignment in which F.P. was

personally involved in 1923 concerned the new harbour at Takoradi in the Gold Coast (Ghana). The port of Sekondi had been the original terminus of the railway opened in 1901 from Kumasi and the Ashanti gold fields, but immediately after the 1914-18 war it was decided to build an entirely new artifical harbour at nearby Takoradi. The scheme involved the construction of two breakwaters having a total length of 2 miles, enclosing a water area of some 220 acres. In excess of 2 million tons of locally quarried granite rubble were required for the breakwaters and this was protected by stone blocks varying in weight between 5 and 12 tons each.

The Governor mistakenly awarded the design and construction contract to a firm of Canadian contractors, not on a 'Package Bid' but on a 'Prime-Cost' basis. By 1922 the quarry had been opened up and a railway laid to the port site, but because of the time taken on these preliminary works, the Canadian contractors were removed. With the aid of a local Government auditor, F.P. was able to show that of the £1.013m expended since the contract was placed, just under £150,000 had gone into the permanent works. He produced one of his masterly reports dealing with the financial, economic, commercial and technical aspects of the situation, estimating the total cost of finishing the work at £3.25m. In due course a new contract was placed with Sir Robert McAlpine & Sons Ltd based on a proper schedule of quantities and prices. Sir Malcolm McAlpine, many years later, admitted to an R. P. T. partner that his firm secured the contract because they strictly adhered to the Consulting Engineer's rates - 'They know more about the site than we do.'

The main wharf, 1500 ft long for handling general cargo and manganese ore provided an alongside depth of 33 ft at low water and was mounted on reinforced concrete cylinders. During the 1939-45 war a bauxite shipment berth was added. The west wharf, 2,150 ft long, dealt with Ghana's main exports – hardwood timber and cocoa – and included the construction of sixteen large warehouses for the storage of cocoa. This stage of the work was completed in 1928, although further major contracts were placed with Taylor Woodrow Construction Ltd. in 1949.

Another example of F.P's prestige in high places arose in November 1926, when the Canadian Government invited him to advise on the selection of a site for a new grain exporting port in Hudson Bay. Ever since the large scale production of wheat had been established on the Canadian prairies at the end of the last century, the producers in Manitoba, Alberta and Saskatchewan had urged their Government to do something about the high costs of transport to the St. Lawrence river ports. It was clear that they would be unable to sell grain in the United Kingdom and Europe unless these costs were reduced. The shipowners and railway companies successfully resisted these pressures for thirty years, but after the 1914-18 War Canada faced severe financial problems and trade restraints, and the matter could no longer be ignored.

Hudson Bay was a logical choice for the new port, because although the distance by sea from both Montreal and Hudson Bay to Liverpool was the same, the latter reduced the inland haulage by between 600 and 800 miles. A start had already been made in 1911 on a new railway branching northwards from the main east to west Winnipeg to Saskatoon line in the direction of Hudson Bay, but the War and the post-war economic depression brought the work to a halt. The problem posed to F.P. was whether the new grain port should be located at Port Nelson or Churchill. Both were small trading posts on the west coast of Hudson Bay, about 600 miles north of the Trans-Canada Railway operated by the then recently formed Canadian National Railways.

F.P. took Ernest Buckton, a recent recruit to RPT's staff, with him, and the two men carried out a detailed investigation of the two sites, often being compelled to take their trial borings through solid ice in the intense cold of the Canadian winter. Finally, Churchill was selected, mainly because it provided a natural harbour where the shipping facilities could be developed in calm water, largely protected from the storms and severe weather conditions encountered in the area. The estimated development costs for Churchill, including an 87 mile extension of the 1911 railway, much of which ran over difficult marsh land, were only half those for Port Nelson. F.P. considered it would take only 3 years to develop Churchill, whereas 6 years would be required for Nelson, where an extensive breakwater would have to be constructed. The Canadian Government accepted his estimates of Canadian dollars 8.45 million for the port installation and dollars 7.54 million for the railway, but R.P.T. was not invited to carry out the detailed design work on the port installation, which perhaps naturally went to a Canadian firm.

* * *

E.J.B. was another remarkable and indomitable character in the Rendel saga.

Ernest James Buckton, (1883-1973), always known to his colleagues as 'E.J.B.', was another remarkable and indomitable character in the Rendel saga. He served his time as a mechanical apprentice and won a Whitworth Exhibition Scholarship, which enabled him to take his B.Sc. (Eng.) at Queen Mary College in London. He was with the old London and India Docks Company when the P.L.A. Consolidation Act of 1908 was passed and he became personally acquainted with F.P. whilst Resident Engineer for their new works programme. During the 1914-18 War he served as a 'Gunner' in France and as a Battery Commander in Gibraltar. During this period he decided that 160 men were the optimum number for an efficient and happy team under his command. In later years, when he was the Senior Partner at R.P.T., the staff were limited to that sacred number and he always boasted that he knew not only the names of every man in the firm, but also those of their wives and children.

After the War, E.J.B. worked as a Deputy Resident Engineer on the harbour extensions at Buenos Aires and took the opportunity of visiting most of the harbours in South America to study breakwaters. He then wisely decided to gain experience with a contractor and joined the colourful figure, Sir John Norton-Griffiths, as Agent and Chief Engineer for the new port at Lobito, in Portuguese West Africa (Angola), the terminus of the Benguella Railway. Later he returned home and joined the Port of London Authority as Chief Draughtsman.

In 1925, when F.P. was called back by the P.L.A. to advise on the big new entrance lock, dry dock and passenger-landing stage at Tilbury, the Authority's Chief Engineer, Asa Binns agreed to release E. J. B. to take charge of R.P.T's Drawing Office. Without doubt, F.P. soon came to the conclusion that Buckton would be his choice for Senior Partner in due course. In 1929, together with Henry John Fereday and Sir Seymour Tritton's son, Julian, he was made a partner in the firm and, upon Sir Seymour's death in 1937 he became Senior Partner.

A job well done invariably leads to another, and following F.P.'s excellent work at Hudson Bay, the Canadian Government instructed the firm to study and report on the provision of a new railway terminal in Montreal. The existing Canadian Pacific Railway terminal was considered to be obsolete, and it was felt that the C. P. R. and the recently formed Canadian National Railway should combine their resources to build a jointly owned terminus on a new site. F.P. confirmed the viability and advantages of this proposal and additionally advocated that the two railway companies should combine their resources in order to completely encircle the island on which Montreal stands, enabling an electrified Rapid Transit suburban railway system to be developed. He proposed that all the transport facilities on the island, including the harbour and tramways, should be combined under the jurisdiction of a single committee, with an independent Chairman, on the lines of the London Passenger Transport Board established in 1933. Unfortunately, no part of the scheme was implemented and the two railway companies remain entirely independent competing entities sixty years later.

The firm became involved with their first major hydro-electric power scheme in 1926. The Perak River, discharging into the Straits of Malacca on the west coast of Malaysia, midway between Kuala Lumpur and Penang had the highest potential source of hydro-electric power in the country. The scheme was first investigated in 1923 by the Civil Engineering Department of Sir W. G. Armstrong Whitworth & Co. Ltd., who had obtained a concession from H.H. The Sultan of Perak for the development and supply of electric power in the State. However, due to difficulties in other parts of the world, they were unable to continue, and the concession was sold.

Lord Elibank, who had been involved with similar schemes in Scotland and a group of friends, formed a new company in London in 1926 and R.P.T. were appointed Consulting Engineers jointly with the specialist Swedish firm of hydro-electric engineers, Vattenbygenadsbyran (VBB) of Stockholm, thus establishing a relationship between the two firms which was to last over 30 years. It was mutually agreed that projects undertaken in the British Empire should be in the name of R.P.T. and that elsewhere the first name should be V.B.B. An old friend of F.P.s, Sir Evan Jones, of Topham, Jones & Railton, who had extensive experience in Malaya and the Strait Settlements, prepared the costs and was later appointed the main contractor.

The scheme comprised a dam of the Ambursen type having a 18.3 metre head, a 27 MW hydro-electric power station on the river at Chendoroh, and an 18 MW steam power generation station at Malim Nawar, together with a system of 110 kV transmission lines. The dam had a discharging capacity of 600,000 cubic feet per second and was fitted with one sector gate measuring 100 ft by 16 ft and three roller gates, 65 ft by 20 ft. The work was complex and difficult, situated in virgin jungle in an area with a high malaria rate. Until the approach road to the site was completed, the only access was by the river. All materials, plant, labour and daily food supplies for the native population of 5,000 workers had to be transported by water from Enggor, some 13 miles down stream from the dam.

The Chief Engineer on site was James Rennie, a member of the great Rennie family of JMR's day. He and his young wife did a splendid job establishing good working and social relationships between the many diverse nationalities engaged on the project. Although neither F.P. nor E.J.B. ever visited the site, Florence Palmer and her daughter Penelope paid a brief visit during a tour of the Far East. The final opening ceremony was performed by the High Commissioner, Sir Cecil Clementi, accompanied by the Sultan of Perak in 1930.

In the Session 1926-1927 F.P. was accorded the great honour of being elected the President of the Institution of Civil Engineers. This meant that for a period of two years he was effectively confined to London, chairing meetings, attending luncheons and dinners and preparing endless speeches. He devoted his Presidential Address to 'Engineering Works in India'. The affairs of R.P.T. were left very much in the capable hands of Sir Seymour Tritton, Ernest Buckton and Harry Fereday during this period. At the conclusion of his term of office F.P.'s wife Florence, presented the Institution with a fine oil painting of him by the artist de Laszlo.

...was accorded the great honour of being elected ...dent of the I.C.E.

Frederick Palmer

A copy of full length painting by de Laszlo painted in 1927 when F.P. was president of the Institution of Civil Engineers.

The reader may be forgiven for having lost sight of the importance of India to the partnership since F.P.'s return home in 1909, and his subsequent pre-occupation with London's docks, the 1914-18 War, Iraq, China, the Gold Coast and Canada. Throughout the War, Sir Seymour Tritton looked after Indian affairs, but increasingly his involvement with the railways concerned mechanical and marine matters. The Inspection Services provided by the firm and the design of specialised vessels such as dredgers, harbour craft, floating cranes and cable laying ships by the Marine Department, were expanded in the post war period. In 1929 the latter department designed and supervised the construction of a large floating dock for the Bengal and North Western Railway. Inspection and the supervision of the acceptance trials of new locomotives and rolling stock for India and other Commonwealth countries became a major activity in the firm. R.P.T.'s Inspection Engineers were frequently visitors to the United States, Germany and Japan, as well as at the private builders' works in the United Kingdom. Important commissions were received also for the design, supervision of the construction and the equipping of new Indian Railway workshops at Lucknow and Perambur.

Bridge building became the responsibility of Harry Fereday (1862-1939). He was born in the 'Black Country' and educated at Wolverhampton Grammar School and Finsbury Technical College. In 1880, he was articled to Robert Braithwaite at the Patent Shaft & Axletree Company and assisted in the manufacture of the Dufferin Bridge over the

Ganges, the first all-steel bridge to be sent to India. After holding several appointments in which he gained valuable experience, including a short period with Colonel R. E. B. Crompton at his Chelmsford works, he joined Sir Alexander Rendel in 1898, and stayed with the firm until his retirement in 1937. He designed his well-known optical stress-recorder in 1917, which has proved to be an important innovation of immense value to engineers for recording the stresses in steel structures under working conditions. Changes in the length of stressed components tilted a mirror reflecting a beam of light, thereby providing a continuous optical or photographic record. Following Sir Alexander's death, he was placed in charge of the firm's steel bridge department and was largely responsible under Sir Robert Gales for many bridges constructed in India. These included the Khushalgarh Cantilever Bridge over the Indus; the magnificent Hardinge Bridge across the Lower Ganges, and today owned by the Bangladesh Railways; the Willingdon and Howrah Bridges over the Hooghly.

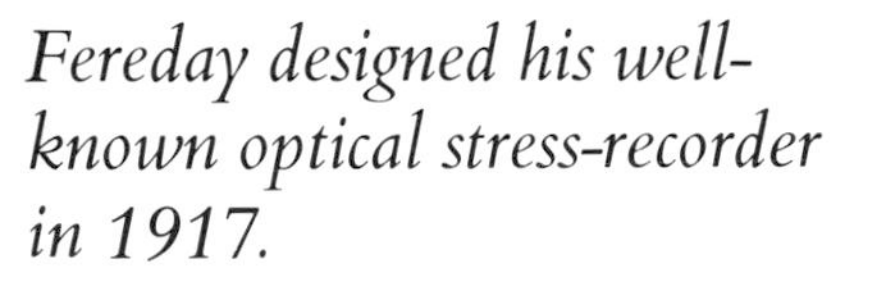

Fereday designed his well-known optical stress-recorder in 1917.

Between the two World Wars Fereday's department was responsible for the following new bridges in India – the Bally Bridge built for the East Indian Railways; the Kekradhar and Purnabhaba Bridges for the East Bengal Railway; the Mehanadiz Kanber and Jhenia Bridges; and the Gorai and Meghnoat Bhaiab Bazar Bridges for the East Bengal and Assam Railways.

In the same period RPT was called in by the Government of Burma to advise them on a number of interesting projects which clearly illustrated the versatility of the Rendel organisation. In 1927, they presented a feasibility study, preliminary designs and cost comparisons for the the Pyagawpu-Yunzalin Water Supply and Power Scheme designed to improve Rangoon's water and power supply. The scheme proposed the supply of 100 gallons of water per day per head for a population of 750,000 people and 27,000 kW installed power from an hydro-electric power station fed from a 12,000 million gallons reservoir having a 400 ft long by 144 ft high dam. The scheme also included a 6000 kW standby steam powered generating station, 111 miles of 66 in diameter steel pipeline with booster stations and purification plant, 130 miles of metre gauge railway, 50 miles of roads, 9 miles of aerial cable way and some 120 miles of 120,000 volt transmission line.

Due to financial stringency, the budget price of £12.25m proved unacceptable and the following year, the alternative Mingaladon tank scheme was put forward jointly with VBB of Stockholm. This necessitated the construction of two tanks or reservoirs covering an area of 9.6 sq. miles, and a 9 miles long, 78 in diameter steel pipeline. The estimated cost showed a reduction of £6m on the earlier scheme.

Harry Fereday (1862-1939) Head of the Steel Bridge Department for many years.

Five years later the firm was commissioned to advise on the possible widening and protection of the Twante Canal, which connected the Rangoon River, via the China Bakir, to the Irrawaddy River. There was concern locally that the width required for the movement of modern vessels might cause excessive scour to the detriment of Rangoon Harbour and the local rice millers' interests. About the same time Harry Fereday's Steel Structures Department were working in conjunction with the Burma Public Works Department. They designed and supervised the fabrication of six pairs of 88 ft long steel trusses for an aqueduct carrying the Salin Canal and a roadway over the Paune Chaung River in the Mimba district.

The firm's final pre-war assignment in Burma dealt with a Government proposal that it should take over control of the village owned embankments on the left bank of the Irrawaddy River. These had been constructed over a period of many years to lower standards than employed by the Government on the right bank of the river. After careful examination of the existing records and comparison with double banked rivers in other parts of the world, R.P.T. submitted a comprehensive technical and economic assessment in which it concluded that double embanking of the river was viable. It emphasised that Government ownership of the embankments on both sides of the river should ensure better maintenance, thus minimising the risks of breaches likely to cause loss of agricultural revenue, which was of paramount importance to the villagers.

* * *

We must now return to the great works closest to Frederick Palmer's heart. As already mentioned, shortly before Lord Devonport's retirement as Chairman of the P.L.A. in 1925, he parted company with his Chief Engineer, Cyril Kirkpatrick, and F.P. was immediately recalled by the new chairman as an Independent Consulting Engineer to supervise the extensive works in hand at the West India Dock and Millwall Dock and at Tilbury. The new Tilbury entrance lock 1,000 ft long by 110 ft wide with a 45 ft depth below high water and the new 750 ft long by 110 ft wide dry dock were finally opened in September 1929 by Lady Ritchie, the wife of the new P.L.A. Chairman. It was agreed that the Orient liner, the *Oransay* with a distinguished party of guests on board chaperoned by F.P., was to make a ceremonial passage through the lock. The plan was that the rolling lift bridge across the lock entrance was to be activated by an ornamental switch on the ship's deck via a shore cable. Unfortunately while the guests were enjoying a sumptuous luncheon below decks as the ship lay off the lock entrance, the cable was accidentally energised. The leaves of the bridge rose majestically without warning, exploding the detonators attached to the roller tracks without a soul in sight aboard the *Oransay.*

The passenger landing stage was opened by the Prime Minister of the day, Mr Ramsey MacDonald, in May 1930. The landing stage for the landing and embarkation of passengers at all states of the tide was buoyed on pontoons. It was 1,142 ft long by 80 ft wide and positioned 270 ft from the shore, access being by means of five hinged bridges. All the facilities required for dealing with the largest and most modern passenger liners were provided, including extensive arrival and departure areas, a customs and baggage hall and a new railway station and associated trackwork. In paying tribute to F.P., the Prime Minister commented, 'his genius is written all over the port.' Later, upon Ramsay MacDonald's recommendation, F.P. was made a Knight Commander of St Michael and St George, 'in recognition of his many major civil engineering works all over the world'. The firm rejoiced and celebrated with him, and during the summer the whole of R.P.T.'s staff, then located at 55, Broadway, in Westminster, were entertained in true F.P. style at Crowhurst Place.

'his genius is written all over the Port'.

In 1929, following a Commons debate, a Royal Commission was appointed to report on the economic aspects of the proposed Channel Tunnel and alternative forms of cross-channel transport. Since the end of the 1914-18 War there had been considerable pressure both inside and outside Parliament to revise this dormant project. A panel of consulting engineers were appointed to advise the Commission on technical and geological questions, including Livesey, Son & Henderson; Mott, Hay & Anderson; and Rendel, Palmer & Tritton. The Royal Commission met twenty-five times, and F.P. personally devoted a great deal of his time to this work. On June 30th, only days after the announcement of his Knighthood, the subject was again put to the vote in the House of Commons and defeated by only 7 votes. The Prime Minister made a statement the following day adding, 'there is no justification for the reversal of the policy

The 20,000 ton Orient liner 'Oronsay' enters the new lock at Tilbury Docks during the opening ceremony in September 1929.

of successive Governments for nearly 50 years in regard to the Channel Tunnel' – and the subject remained closed until after the Second War.

The remaining years of Sir Frederick Palmer's life were mainly devoted to assignments in the Levant. In the years preceeding the 1914-18 War the initiative for developments in Turkish dominated Syria and Palestine rested with the Germans. In 1901, they built the 1,065mm gauge Hedjaz Railway from Damascus to Medina in Saudi Arabia, the Mecca pilgrims line which received the destructive attention of Lawrence of Arabia during the War. In 1905, a branch line was built from Damascus to Haifa for the export of Syrian grain. Sheltered from south-westerly gales by Mount Carmel, Haifa provided an excellent deep water safe anchorage.

When Palestine became a British Mandated Territory in 1918, the government proposed a standard gauge railway from Ismailia in Egypt, through Gaza, along the coast to Haifa. Although, only the northern part was completed, this gave a new impetus to the development of Haifa as a modern port. The Admiralty also decided to establish a naval base in the Bay as part of its improved defences in the Eastern Mediterranean.

Sir Frederick Palmer proposed a 300 acre harbour protected by a one-and-a-half mile long breakwater with a 40 ft depth of water at the round-head and a half mile long lee breakwater forming 600 ft wide entrance. The 1,400 ft long wharves had a 31 ft depth of water alongside and the 900 ft long lighterage wharves had 16 ft of water. Stern-on moorings for a dozen or more ships were provided along the main breakwater, and a 25 acres oil dock was proposed along the lee breakwater. Graded rubble was used for the construction of the breakwaters having wave-breaker blocks of 7 to 15 tons on the seaward face and a concrete block parapet. Some 1.8 million cubic metres of material dredged from the harbour area was used to reclaim 100 acres of land now used for urban development. R.P.T. carried out the initial survey and investigations and produced the contract drawings, but the work was carried out by the Haifa Harbour Works Department, using direct labour under R.P.T.'s technical supervision. The work was completed in 1933 at a cost of £2m.

The Germans cherished the notion of a great railway from Berlin to Baghdad.

Ever since the 1880's the Germans had cherished the notion of a great railway from Berlin to Baghdad. Georg von Siemens and the Deutsche Bank were the promoters of the line from Istanbul across Turkey, and this project was a constant source of political trouble in the decade before 1914, and contributed materially to the tensions leading up to the 1914-18 War.

After the war, the British Government instructed the Crown Agents to investigate the practicality of a railway from Haifa to Baghdad and in 1931 RPT was appointed to carry out the initial survey. The proposed 685 miles long route from Haifa followed the historic plains of Armageddon and Jezreel before descending to cross the River Jordan at a point 870 ft below sea level. The route then ascended the steep hills of ancient Gilead on the Jordan plateau 2,350 ft above sea-level. It then crossed the Iraqi desert to the River Euphrates and followed the river for 50 miles before heading east to its terminus at Baghdad on the River Tigris. Six survey parties were organised by the firm and the field work was completed in twelve months. Subsequently, designs and estimates were prepared and a report was submitted, but for all practical purposes that was the end of the matter.

In the early 1930's there was a lot of discussion about the rebuilding of several of the great Thames Bridges in Central London to cope with the growing traffic problems. Undoubtedly, it was a great ambition of Sir Frederick's that the firm should be involved in these projects from their inception. They were appointed consultants to the promoters of a new Charing Cross Road Bridge, which involved the removal of the old Hungerford Railway Bridge, but Parliamentary Approval was denied and the project failed. He did, however, secure the commission for the demolition and rebuilding of Rennie's old Waterloo Bridge, but that story belongs to the next chapter.

Throughout this stage of his life, Sir Frederick had become plagued with a progressive form of arthritis. Increasingly, his lameness made him dependent upon others, something he found extremely hard to accept. He slept badly, and although surrounded by everything a man could desire, his life became far from a paradise. Upon the advice of his doctor, the celebrated Lord Dawson of Penn, King George's Physician, he went to a spa in Czechoslovakia for treatment, but this proved to be of little benefit. The end came in his 73rd year, at noon on Saturday, 7th April 1934. In accordance with his own wishes, he was buried in the little parish churchyard of St. George at Crowhurst. A week later a large and distinguished gathering attended a memorial service held at Westminster Abbey. His passing marked the end of an epoch, for he was only the third Senior Partner in the firm spanning a period of no less than ninety-six years. Things were never quite the same again.

Things were never quite the same again.

7
New Blood Takes Over 1935-1945

Upon Sir Frederick Palmer's death, the senior partner's mantle was immediately taken up by Sir Seymour Tritton. He was assisted by Sir Robert Gales, Ernest Buckton, Harry Fereday, and his son Julian Seymour Tritton as partners. Sir Seymour was over seventy-four upon his accession and was content to remain the figurehead, although he still retained the appointment as Consulting Engineer to the Indian Government. Sir Robert Gales was also over seventy and, apart from the Jinja Bridge in Uganda, the Falluja Road Bridge over the Euphrates, and the delicate negotiations with RPT's competitors, Livesey and Henderson, in connection with the Lower Zambesi Bridge project, he took little part in the day-to-day running of the business. The triumvirate – Ernest Buckton, Harry Fereday and Julian Tritton – carried most of the responsibility for the continued success of the business. Buckton was the Civil Engineer, Fereday dealt almost exclusively with steel structures and Julian Tritton was the Mechanical Engineer who concerned himself mainly with railway related matters, the inspection services and the growing Marine Department.

Sir Robert Gales's Jinja Bridge was built across the River Nile on the north shore of Lake Victoria where the river drains out of the lake over the Ripon Falls. It was commissioned by the Kenya and Uganda Railways and Harbour Board as part of its extension to Kampala. The single line, 3 ft 6 ins gauge railway was carried on the top deck and a 20 ft wide roadway used the lower deck, which was suspended from the main arch span. The main span was spandril braced and 260 ft long between hinges with 100 ft long approach spans on either side. The bridge was substantially modified in 1949 when the nearby Owen Falls Dam was built.

Sir Alexander Henderson, the 1st Lord Farringdon, had already amassed a considerable fortune as a railway promoter when he became aware of the need for a new railway in East Africa, to connect Zambia and the Nyasaland Protectorate (now Malawi) with the Mozambique port at Beira. In 1902 he financed the formation of the British Central African Company to build and operate the railway.

At the end of the 1914-18 War, Lord Farringdon's youngest brother, Sir Brodie Henderson, returned to the family-owned firm of consulting engineers, Livesey, Son & Henderson. He was a first-class engineer and administrator, and during the war he became Deputy Director General

Julian Tritton.

Jinja Railway Bridge, Uganda.

of Transportation in France with particular reference to railways. In this capacity he had close contacts with Sir Seymour Tritton. After the war he supervised the completion of the new docks at Buenos Aires, and succeeded Sir Frederick Palmer as President of the Institution of Civil Engineers in 1928.

It was decided that a railway bridge should be built across the Lower Zambesi, between Dona Ana and Sena in Mozambique, the lowest point at which the river could be crossed. The bridge is 2.24 miles long and comprises 33 main arched spans, each 253 ft long, and for many years it remained the longest bridge in the world. RPT carried out the preliminary survey and investigation work, chose the site, designed the arched spans, produced the contract drawings and supervised the fabrication work. Livesey & Henderson, under Sir Brodie Henderson's leadership, retained the site supervision to themselves and presented a paper before the Institution of Civil Engineers, claiming most of the credit. Be that as it may, the bridge remained a great triumph for British engineering for many years. Since the establishment of the People's Republic of Mozambique in 1967, the bridge has been owned and maintained by the State Railways, C.F.M.

... the bridge remained a great triumph for British engineering.

The Lower Zambesi Railway Bridge, Mozambique 2.24 miles long having 33 main arched spans each 253 feet long, and for many years the longest bridge in the world.

The Ava Bridge across the Irrawaddy River in Burma completed in 1934 carrying the metre gauge railway and roadways on either side.

In the same period the firm designed and supervised the construction of a not dissimilar bridge for the Government of Burma across the Irrawaddy River near Mandalay on the metre gauge Rangoon to Mandalay railway Some 3,760 ft long, the bridge had nine arched river spans each 350 ft long, one of 250 ft long, and six approach spans each 60 ft long. Thirteen feet wide roadways were cantilevered on either side of the single track railway. Two of the main river spans were destroyed during the 1939-45 War and subsequently replaced prior to Burma's withdrawal from the Commonwealth and virtual isolation.

As a preliminary to the extensive road works carried out in Fiji in the 1970's, RPT was commissioned to build a through truss road bridge across the River Rewa, built on 80 ft deep caisson foundations. Completed in 1934, the merit of this work was not so much the physical design of the bridge, as the organisation and competence which enabled the work to be efficiently executed on the other side of the world, on a comparatively small Pacific Island.

No Civil Engineering project was too large, too remote or too difficult.

We have reached the stage in our story when it may be said that the world was RPT's oyster. No Civil Engineering project was too large, too remote or too difficult for them.

* * *

At home in the same period, the Partnership was retained by West Ham Borough Council in London to design, prepare contract drawings and tenders, and to supervise the construction of Silvertown Way and Silvertown By-pass. Since JMR's days in Devon and Cornwall working on the Turnpike roads a 100 years earlier, the firm had had little involvement with new road construction. In the 1920's it was suggested that toll roads should be re-introduced, and RPT was associated with a scheme to rebuild the London to Brighton road as a toll road. Life has the habit of turning a full circle, and motorway toll roads are today common-place in many countries, but their introduction into the dense network of this country, attractive though this might be, would present many serious problems.

The four-lane Silvertown Way forms part of a scheme designed to improve the approaches to the Royal Docks. It includes a three-quarters of a mile long viaduct with a spur road giving access to the Royal Albert and King George V Docks. The reinforced concrete viaduct incorporates unique highly stressed concrete hinges which were the first of the many

The one and a half mile long Silvertown Way forming part of the improvement to the approaches to the Royal Docks, London, completed in 1934.

innovations which the then young engineer, John Cuerel, was to introduce in advancing concrete technology. Although most of the docks are now disused, these roads and the viaduct – a very early urban elevated roadway – provided essential arterial links in the redevelopment now taking place.

The remaining major oversesas projects completed during Sir Seymour Tritton's tenure as Senior Partner were the improvements to the Lighter Port at Jaffa in Palestine, and hydro-electric schemes in Australia and Venezuela. The success of Sir Frederick Palmer's work at Haifa encouraged the Government to proceed with the scheme at Jaffa. A new 3,000 ft sea wall and breakwater was authorised, consisting of a crest of two lines of precast dovetailed concrete blocks with a mass concrete core anchored to the natural reef. New 1,800 ft long deep water quays and associated facilities resulted in an upgraded port. The work was completed in 1937, RPT seconding key staff to supervise a direct labour contract.

The Australian Victoria State Electricity Commission were responsible for a scheme designed to improve the supply of electricity to Melbourne. RPT were appointed consulting engineers to the Commission in

Perak River Hydro-electric Power Scheme, Malaysia, designed by RPT in association with their Swedish colleagues VBB of Stockholm.

In the 1930s Rendel Palmer & Tritton was responsible for major improvements to the Lighter Port at Jaffa. The work included about 3000 feet of sea-wall and breakwater and 1800 feet of quays, lighter moorings and a slipway.

association with their Swedish colleagues VBB of Stockholm. The project involved the development of a scheme to utilize the water power of the Kiewa River, necessitating three 125 high rock-filled and earth, concrete arch dams, a four miles long diversion tunnel, and four power plants having a combined head of 4,065 ft producing a total installed capacity of 117 MW. The total cost was £4.5 million.

Concurrently, the German firm Siemens-Schuckertwerke commissioned RPT to report on a proposed two-part hydro-electric scheme on the River Cura, forty miles south of Caracas in Venezuela, designed to produce a total installed capacity of 10,300 KVA. The first generator station, situated at the confluence of the rivers Rio Carmen and Rio Guan, was fed by water stored in two reservoirs 4270 ft above sea-level. The second generator station was fed from a dam diverting water into an 11,000 ft long steel pipe immediately downstream on the River Cura.

The Snowy Mountain Scheme initiated by the Government of New South Wales, remains one of the greatest civil engineering works ever undertaken in Australia. It was first proposed in 1915 and was given form by W. Corin, the then Chief Electrical Engineer of the Public Works Department, in 1918. He proposed a huge integrated hydro-electric and steam power generator scheme and the electrification of the railways radiating from Sydney to the growing mining and industrial towns of Newcastle, Lithgow, Goulburn and Nowra. Due to financial constraints the scheme lay dormant until the mid-1930's, when RPT and VBB were commissioned to report on recent developments in power generation and the electrification of railways. Their report led ultimately to the massive 2500 MW scheme completed in the 1960's, although neither firm was involved in the project as finally developed.

The first important bridge designed by RPT in London was the Chelsea Suspension Bridge.

The first important bridge to be designed and built by RPT in London was the Chelsea Suspension Bridge crossing the River Thames between Chelsea and Battersea. The work of replacing the original suspension bridge on the same site was commissioned by the London County Council in 1932 and completed in 1937. The bridge was the first self-anchored suspension bridge to be built in England. The main span is 352 ft

long and the side spans are each 174 ft long. Each stiffening girder is made up of two steel plate girders 8 ft 10 ins in depth supporting a concrete deck spanning between the girders. The four towers, carried on transverse hinges which provide lateral stability, are supported on piers found in cofferdams on London clay 30 ft below the river bed level.

The most difficult problem with suspension bridges is often the provision of adequate end anchorages for the cables. Dependent upon the topography and nature of the ground, extensive and complex provision may be involved. RPT's team of bridge designers, led by Harry Fereday, solved the problem by taking the tensile forces back into the deck of the bridge, rendering it 'self-anchored'. The credit for employing this solution in the Chelsea Bridge lies largely with J. R. H. (Joe) Otter, then a trainee engineer with the firm. He based his designs on the 1,040 ft long centre span suspension bridge across the Rhine connecting Cologne and Mulheim. The story is told that Joe, a former pupil at Westminster School, had a particular terror of his German Master. In the early stages of his career as an engineer he put his school German to good use and translated an article describing the design and construction of the

self anchored Chelsea Suspension Bridge ss the Thames. Main span 352 feet , side spans each 173 feet and iageway 40 feet wide with ilevered footways. Completed in 7.

H. (Joe) Otter, B.Sc., M.I.C.E. liant engineer, mathematician and puter programmer. Involved with design of the Howrah, Chelsea Waterloo Bridges. Chief engineer manager Merz, Rendel, Vatten kistan).

Mulheim Bridge. Harry Fereday listened to his pupil and allowed him to develop the idea. This was probably the first and only occasion in which a trainee has been entrusted with the design of a major bridge. The idea of subjecting a girder to compressive force throughout its length was subsequently further developed in order to eliminate tension from materials such as concrete, which were inherently weak in resisting it.

Joe Otter was a most remarkable man in many ways. He was a brilliant mathematician, he excelled at chess and he loved classical music. In 1927 he joined RPT, and after preliminary training became wholly engaged upon the Chelsea Bridge project and the demolition of Rennie's old Waterloo Bridge across the Thames. Later he went to India as the Assistant Resident Engineer on the great Howrah Bridge project in Calcutta, which was completed in 1942. He then joined Merz & McLellan in India engaged upon the construction of power station cooling water works and a new dry dock for the Indian Navy.

In 1955 he returned home and re-joined RPT and was made a partner in the following year. He assumed responsibility for a number of road and road bridge projects, but he recognised the potential of the computer, then becoming available as a working tool, and, almost intuitively, wrote a wide variety of programmes for modelling complex and time related designs, and even accountancy. His brilliance in this field was widely acknowledged in the engineering and academic world. Following the collapse of the Second Narrows Bridge at Vancouver during its construction in 1985, he was appointed an adviser to the Royal Commission investigating the affair.

Francis Irwin-Childs, a former senior partner, tells the story that when physical models of the proposed Thames Barrier were being evaluated, it was found that the model did not extend sufficiently far to the east. Joe Otter took the problem home one week-end and returned with a mathematical model solution based on first order finite equations. His solution not only confirmed the physical model but also effectively extended it some 80 miles into the North Sea. Later Otter and A. S. Day developed the technique to provide solutions to the design of prestressed concrete pressure vessels to nuclear power stations, arch dams, thin shells and heat transfer. His premature death in 1969 was a sad set-back to further development of his appreciation of the power of computer technology.

Rennie's original elegant Waterloo Bridge was officially opened by the Prince Regent in 1817. In 1884, it was found necessary to protect the pier foundations against scour, and aprons of concrete enclosed by sheet piling were added. In 1924, serious settlement occured in Pier 5, and the bridge was temporarily closed. Two years later a Royal Commission recommended that the bridge should be reconditioned and widened to accommodate four lanes of traffic. When Parliament rejected the scheme for a bridge at Charing Cross in 1932, the L.C.C. resolved to build a new Waterloo Bridge. Four years were occupied by in-fighting between the Government and the L.C.C., the former refusing to finance a new bridge. However, in 1936, Parliament did approve an item in the Council's Money Bill for the reconstruction of Waterloo Bridge, which enabled the L.C.C. to meet the cost of demolition and reconstruction out of borrowed money. In other words, the new bridge was to be constructed 'on the rates'.

The hospitality extended to Sir George Humphreys, the Chief Engineer of the L.C.C., by Sir Frederick and Lady Palmer at their Crowhurst Place week-end house parties paid dividends. Despite fierce competition from other consulting engineers, RPT was appointed to supervise the demolition, carry out site investigations and detailed designs for the new bridge, prepare the contract documents, report on the tenders and supervise the construction work. The famous architect, Sir Giles Gilbert Scott was retained to safeguard the aesthetic appearance of the new bridge. E.J. Buckton assumed responsibility for planning the work and John Cuerel carried out the detailed design work.

John Cuerel (1903-1972) was educated at Weymouth Grammar School and graduated at Bristol University. Initially, he spent some time with a contractor engaged on tunnelling and railway work before joining the Port of London Authority working under F.P.'s supervision on the Tilbury Dock improvements. He joined RPT in 1929 and quickly assumed responsibility for reinforced concrete bridge work, marine structures, power stations and other heavy foundation work. Like Joe Otter, he had an instinctive flair for design solutions. He was made a partner in 1946 and became the senior partner in 1961, a position he held until his retirement in 1966. He was a past President of the Reinforced Concrete Association. Waterloo Bridge, more than any of his works, remains a fitting monument to his memory.

The final design of the bridge developed in collaboration with the architect Sir Giles Gilbert Scott was not achieved without great difficulty. A deck type structure was needed in order to preserve the view down river with St. Paul's in the background. This and the navigational headroom required resulted in a very restricted construction depth. To avoid a tunnel effect the soffit was required to be divided into two 'arches'. The resulting design problems were solved by Cuerel with great ingenuity.

The bridge comprises five main spans made up of twin reinforced concrete box girders, each continuous over two spans with projecting cantilevers to support the suspended centre span. This arrangement gives three clear spans of 250 ft each and two outer spans of 240 ft having short cantilevers at their outer ends.

The slender lines required a high concentration of reinforcement. Butt welding, then infrequently used, was used extensively to avoid the congestion of conventionally lapped bars. Medium high tensile steel bars joined with threaded couplers and contained in tubes were steam heated before nuts at their ends were tightened. On cooling, the bars were thus prestressed and they were then protected by the grouting of the tubes. Closely spaced grids of reinforcement, welded at the laps, controlled surface cracking more effectively than hitherto. Even today the virtual total absence of cracking or rust staining is testimony to the high standards of workmanship required by Cuerel, and which are similarly evident in the appearance of all his other projects.

Movements due to deflexion and temperature were simply dealt with by supporting the deck on thin flexible concrete bearing walls fixed into the foundations at the base of the piers. Shell piers around the bearing walls protected them from ship collisions.

The elegance of the free flowing clean design was preserved when it was decided to leave in place the simple tubular handrail, erected as a temporary war time economy, in place of the ornate railing of the original design and also to omit the arches over the approaches as earlier proposed by Sir Giles.

John Cuerel, Senior Partner Rendel Palmer & Tritton 1961-1966. Responsible for the engineering design and construction of London's Waterloo Bridge.

* * *

Major changes took place in the partnership in 1937, the 99th year of the firm's existence. Sir Seymour Tritton died on November 21st and was

Waterloo Bridge, London. John Cuerel's masterpiece designed and constructed in association with the distinguished architect Sir Gilbert Scott, R. A. Completed in 1942.

immediately succeeded by Ernest Buckton as senior partner. Sir Robert Gales and Harry Fereday retired, and Ralph Strick, a Welshman, known to Sir Fredrick Palmer since the days of the ill-fated Shatt-Al-Arab project, was brought into the partnership by Ernest Buckton.

Strick was born at Neath in 1886 and studied engineering at Swansea Technical College, where he obtained a B.Sc. degree. In 1907 he joined Topham, Jones & Railton, and for the next nine years was employed mainly on the construction of the King's Dock, Swansea. During the 1914-18 War he held a commission in the Royal Engineers and was one of three officers responsible for devising the plan to block Zeebrugge harbour. He ended the war in Mesopotamia, and stayed on to become the Port Engineer of Basrah. Following the dispute between Sir Frederick Palmer and the Port General Manager, Colonel John Ward, he resigned. From docks and waterways, he turned to railways and accepted the appointment as Bridge Engineer on the Bengal-Nagpur Railway. Perhaps his greatest achievement during this stage of his career was the doubling of the Rupnarain Bridge without interrupting the railway's single line working. The project involved the erection of seven 300 ft long and four 100 ft long spans. He also carried out a major bridge strengthening programme involving the removal, replacement and conversion of bridge spans on a large scale. By immediately appointing him a partner, the RPT partners obviously considered he was the right person to replace Sir Robert Gales as the firm's man in India. Of special concern at this time was the satisfactory completion of the £2.5 million Howrah Bridge being built across the Hooghly River between Calcutta and Howrah. Their confidence was well placed and ten years later Strick became the senior partner of the firm.

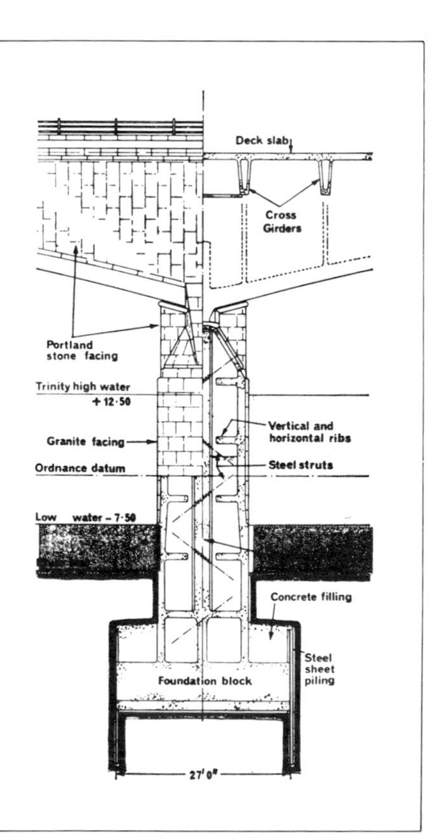

Cross section of a pier of John Cuerel's Waterloo Bridge.

Since the end of the nineteenth century, engineers have sought a means of replacing the old floating bridge at Howrah, opened in 1874 and referred to in Chapter 5. It was not until 1933 that, with the advent of high tensile steel, it became financially viable to build a fixed bridge. The design adopted is a cantilever bridge with a river span of 1500 ft similar to that recommended by Sir Frederick Palmer in 1922. The anchor arms each have a 325 ft span, the carriageway width is 71 ft wide footways.

Ernest Bateson was given responsibility for the foundations. John Cuerel designed the reinforced concrete monoliths with square dredging holes, forming the piers and abutments. The main piers are 181 ft 6 ins long by 81 ft 6 ins wide, the Calcutta piers being sunk to a depth of 105 ft. Probably for the first time, the long settlement in the soft alluvium was calculated from tests using a consolidation cell manufactured at the bridge site. Recent measurements show the actual settlement to be within ½ inch of the calculated amount. F.M. Fuller assisted by Arnaud Toms prepared most of the calculations for the steelwork, which were filed in some 200 thick volumes and carefully indexed. No Code of Practice for the use of high tensile steel in bridge construction existed at the time and the firm had to devise its own standards, which subsequently formed the basis for the British Standard. A novel feature was the prestressing of members during erection in order to eliminate secondary stresses under basic loading conditions.

Ralph Strick, Senior Partner of Rendel Palmer & Tritton 1947-1961.

RPT's first involvement with Civil Aviation occurred in 1937. They were invited by the Air Ministry to investigate and report upon the suitability of a site at Bradwell Quay in Essex on the River Blackwater, as a base for flying-boats used on the Commonwealth routes. They carried out general investigations, trial borings and detailed estimates for the first stage of the development, but the scheme was frustrated by the war. During the war, designs were prepared for a similar base on the River Shannon in Eire. The debate as to whether flying-boats or conventional aircraft should be used on trans-Atlantic routes continued until the 1950's.

RPT celebrated their centenary with a dinner on 19th December 1938.

RPT celebrated their centenary with a dinner at the Dorchester Hotel on 19th December 1938. Over 500 guests attended, including representatives of the technical and professional institutions, home and overseas railways, Government Departments, the Crown Agents for the Colonies and many others associated with the engineering profession in all its branches. In proposing a toast to 'The Guests', Julian Tritton paid a special tribute to Lady Florence Palmer who was present. Replying, Mr W. J. E. Binnie, the President of the Institution of Civil Engineers, traced the role of the engineer in history, adding:

> It was the engineers' job to put into practical terms, or make useful to the community, what the scientist had discovered. In Ancient Rome, the engineer was a person of great importance, and might even be proceeded by lictors to clear the way for him as he proceeded to his office. He had traced contractors back to a very early period and was not sure the Pyramids were not built by contract. Cicero had said contracting was the best way of making money quickly and honestly.

Lt. Colonel Edwin Kitson Clark, then head of the great Leeds based locomotive building firm, then addressed the gathering, tracing his own firm's long association with RPT, when his grandfather had worked closely with Sir Alexander Rendel. He referred to the many happy years he himself had enjoyed an intimate relationship with Sir Seymour Tritton, adding that there had never been any question between them that had not been solved at once in a sympathetic and sensible way. He said that he had always thought Kipling was much to be thanked for giving engineers a very great idea of the dignity and importance of their work. There was a nobility in their profession that made them missionaries to the world, and with that nobility the firm of Rendel, Palmer & Tritton had a very great deal to do.

In his reply to the toasts, Ernest Buckton referred to matters which fifty years later still remain unresolved and the subject of frequent debate. He urged that large engineering structures should be put in hand in slump periods so as to be ready for the next boom. Discussing the future he said

The high tensile steel bridge across the River Hooghly connecting Calcutta and Howrah and having a 1500 feet long river span. Completed in 1942.

that he considered Communism and Fascism had done great work, for they had made democracy healthier than ever. He suggested Capitalism was the problem of the future for democracy. Money and labour were both necessary to produce profits and both where possible should share in profits. He stated his belief that democracy would be healthier with an extension of the profit sharing system for labour, and that this was a matter in which RPT was trying to be helpful in a material way.

Only weeks before RPT celebrated its centenary at the Dorchester Hotel, Prime Minister Chamberlain had signed the abortive Munich Agreement, and inspite of his promise of 'Peace in our time', the war clouds gathered. Re-armament went into top gear, the Germans annexed Czechoslovakia in March 1939, and on 3rd September 1939, following the invasion of Poland, Great Britain and France declared war on Germany.

Work on Waterloo Bridge and the Hooghly River Bridge in India continued under Fereday and Cuerel's supervision, but increasingly members of the RPT team were recruited or compulsorily mobilized for work of national importance. Julian Tritton and Ernest Bateson remained with the firm undertaking important work for the Ministry of Aircraft Production on the design and construction of radar installations. The firm was also commissioned to supervise the construction of forty-nine satellite landing grounds (S.L.G.'s) scattered throughout the United Kingdom which never appeared on aeronautical charts. They were intended for the storage of the ever increasing number of aircraft coming off the production lines, safe from the eyes of the Luftwaffe. Parks, golf links, race courses and farms were requisitioned overnight and Spitfires, Hurricanes and Wellingtons replaced pheasants in woodland glades.

The denuding of the established firms of consulting engineers of their key staff became so serious that Ralph Strick felt compelled to protest publicly. He asserted that the great expansion of government departments and the formation of new ones at the expense of the established private firms previously performing the same functions, was wrong and damaging to the public interest in the long term. He pointed out that inevitably time would be wasted whilst incoherent collections of individuals were settling down into a semblance of an efficient team, and that by crippling the long established private organisations, essential reconstruction would be seriously retarded when victory was achieved.

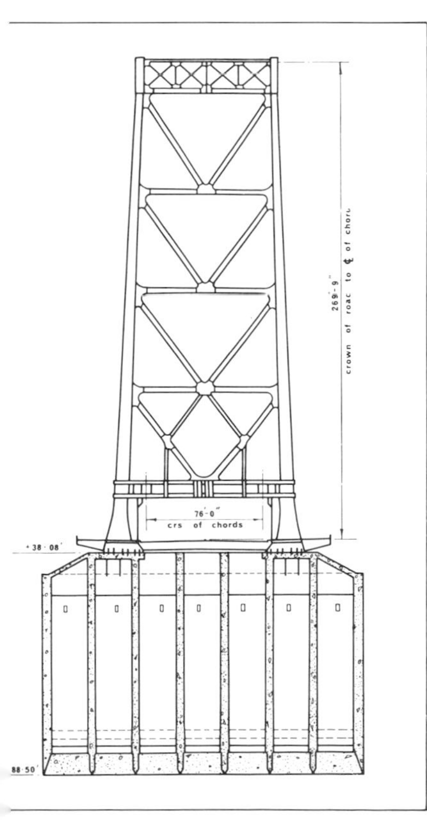

Howrah Bridge across the Hooghly River section showing the 105 feet deep Calcutta pier.

8
Bridges, Roads, Railways, Airports & Water Resources

Within a year of the final defeat of Nazi Germany in May 1945 and the unconditional surrender of Japan three months later, Ernest Buckton completed his re-organisation of the partnership, and laid foundations which enabled the firm to take full advantage of the many worldwide opportunities which arose in the post war period. Alfred John Clark, Frederick Alfred Greaves (no relation to JMR's assistant in the 1850's), Ernest Bateson, and Sir Frederick Palmer's son, John Palmer, were made partners and other key personnel returned to the fold, several having rendered distinguished war service during their absence.

Buckton endeavoured to retain the old traditions of the Raj, which had been so characteristic of the partnership since Sir Alexander Rendel's day, but he realised that the introduction of new technologies and commercial, economic and political change were inevitable. Several highly qualified younger men were recruited, and they soon made their mark in specialist fields as these changes began to take effect. Increased competition was expected from both home and abroad. But Buckton ensured that the highest professional standards, always a cornerstone of his predecessors' attitude to the business, were maintained.

One of the firm's first major post war developments occurred in 1948, following the granting of Indian independence. RPT formed a consortium with Merz and McLellan, Consulting Electrical Engineers, and their pre-war Swedish associates, Vattenbyggnadsbyran (VBB), who were experts in the design and construction of hydro-electric, water and sewage schemes. The new company was styled Merz Rendel Vatten (Pakistan) having its Registered Office at RPT's headquarters at 125, Victoria Street, London, and branch offices were established in Karachi and Lahore. Joe Otter, who was later to be made a Partner, worked in the consortium.

Within two years the Government of Pakistan, and the provincial Governments of Punjab, the North West Frontier, Sind and East Bengal, later to become Bangladesh, had entrusted the new company with several major projects which included hydro-electric schemes, steam and diesel-engine-driven power stations, the enlargement of the port of Chittagong, a new dry dock at Karachi and the preparation of a master plan for Greater Karachi to cater for an anticipated population of 3 million people.

E.J. Buckton — Senior Partner, Rendel Palmer & Tritton, 1937-1947.

By 1950 over seventy per cent of the firm's work had been commissioned by overseas clients.

By 1950, over seventy per cent of the firm's work had been commissioned by overseas clients and during the five years following the end of the war, the staff was progressively increased to a total of 480 employees.

The Railway and Marine Department, including the activities of the Inspecting Engineers, was under Julian Tritton's control and supervision. Alfred Clark, Alfred Greaves and John Palmer shared responsibility for the Harbours and Roads Department. Ernest Bateson was in charge of steel bridges and steel structures and John Cuerel headed the Structural Concrete Department which dealt with bridges, jetties, quays, foundations and other reinforced concrete structures. The Accounts and General Administration function was the responsibility of Harry Sharpe and upon his retirement in 1966, he was succeeded by John Nicholls, who continued as Financial Controller and Secretary to the Partnership until its termination.

The organisation of the office was based on section leaders in charge of projects or parts of projects assisted by teams of people of various qualifications and degrees of seniority. Frequently, these teams included consultants and others qualified in specialised disciplines, often quite distinct and different from civil engineering, but which formed an essential part of the comprehensive service offered by the firm. A barrister was also attached to the firm to check the conditions of contract and the contents of letters which might unwittingly contain some contentious legal point unseen by the layman.

So vast was the number and diversity of the projects undertaken by the firm in the forty years following the ending of the 1939-45 War, it is impossible to deal with them individually in the space available. We must content ourselves by briefly considering the types of work and noting the changes and innovations which they brought about.

It has often been said that such is the rate of technological progress, three-quarters of what we take for granted today did not exist fifty years ago, and that the same will be true in twenty-five years time. Let us therefore take a brief glimpse at the state of the art as it existed at the end of the war.

Both in situ and precast reinforced concrete were well established and prestressed members and structures were just beginning to make their appearance. Great Britain had lagged behind the United States and some European countries in this field, but during the war it was shown that the economic factors for using prestressed concrete were of considerable importance. There was considerable controversy in some fields between the advocates of reinforced concrete and structural steel, especially in situations where fire resistance was important, as in transit sheds, warehouses and similar buildings. Heat distortion effects on steel frames often led to progressive collapse, whereas reinforced concrete framed structures were much less subject to these problems. The protection of steel frames later improved but the debate continues to the present day as changing costs alter the economic balance.

Welding received active consideration during the war as an alternative to riveting. Problems had arisen with welded pressure vessels and with ships' hulls and there existed in many engineers' minds a nervousness regarding the reliability of welded compared with riveted joints. Little had been laid down in the nature of standard procedures for the guidance of designers and operatives and there was still a large element of empiricism. Consulting engineers, including RPT, played an important rôle in developing standards and restoring confidence.

RPT played an important rôle in developing standards.

Timber had for long been regarded as a 'friendly' material for marine jetties and similar structures. Imported hardwood was still cheap, and readily available in many overseas countries. It was easy to erect and errors in construction or damage caused by ship collision could be easily rectified by carpenters. Unfortunately, it was subject to attack by marine borers, and even when suitably impregnated had a comparatively short life. The introduction of reinforced concrete for marine work was accelerated by the development of aids for bringing large vessels more gently alongside. The development of dolphins and spring or cushioned fenders greatly reduced the fear of damage and in this field RPT engineers played a major rôle, some of the designs being patented.

At the end of the war, Soil Mechanics was a rapidly developing science led by Dr Karl Terzaghi in the United States, but it had not reached the stage when it had been accepted by engineers in practice and empirical methods persisted. RPT was one of the first major British civil engineering consultancies to establish its own Geotechnical Department which was initially headed by Kenneth Ainscow. He eventually became a partner, succeeding Joe Otter in 1968. Undoubtedly, the ancient Chinese and the Egyptian Pharaohs and later the Romans possessed some appreciation of soil and rock mechanics, albeit at a subconscious level. Today, no civil engineer would embark upon the design and construction of foundations, underground excavations or rock face structures without first obtaining a detailed appraisal using modern geotechnical techniques. These include trial borings, mapping, the laboratory testing of samples and the computerisation of the results. To-day, in the preparation of feasibility studies and preliminary designs, during the detailed design stage, and during the supervision of construction, and as a part of the post-construction services offered by the firm, geotechnology plays a vital role.

Kenneth Ainscow – Partner 1968-76.

In the early post-war years there were no electric typewriters or word processors, no photocopying machines or microfilms and, of special interest to engineers, there were no computers or electronic calculators. Most design was done by slide rule or hand calculator and, where particular accuracy was essential, long and tedious calculations using logarithms had to be made. The telex machine was in its infancy and most urgent overseas communications were sent by cable. International telephone calls were costly and unreliable, often requiring booking many hours in advance, making communication with overseas staff difficult. Airmail postal services were just coming into operation. Inevitably, life moved at a slower pace forty years ago, and in retrospect the one great advantage was that most people were compelled to travel by sea when making overseas visits. Long distance air services were minimal and not very reliable, and the long sea voyages provided normally busy executives with an enforced opportunity of relaxing, able to enjoy undisturbed thought, catching up with their reading, and preparing written reports in a congenial, health giving atmosphere.

In his Presidential Address given before the Institution of Civil Engineers in January 1852, the founder of the firm remarked,'The principal works in which, in these days, the aid of the civil engineer is required, are railways, harbours, docks, bridges, the improvement of rivers, the formation of canals, the drainage of lowlands, the supply of water to towns and their sewerage.' A hundred years later all these interests remained in the partnership's portfolio, and as the scope of the civil engineer has expanded with changing technology, JMR would have today added, had he been alive, highways, the production of electrical power, oil exploration and oil pipelines, coast and flood protection works and airfields. He would have been amongst the first to applaud and utilize the benefits of the computer for structural, hydraulic, and tidal analysis and for earthwork calculations. He would have been immensely proud of the sophisticated work now carried out by the firm in dealing with the economic and

JMR would have been immensely proud of the sophisticated work now carried out by the firm.

technical problems of urban, industrial and dockland re-development, and with the preparation of transportation and highway studies. He would have been fascinated by the work of Joe Otter and others in the preparation of mathematical models for solving complex problems, and by the use of aerial surveys for route alignment. One suspects that today he would have been amongst the pioneers, actively applying his mind to the challenge of wave energy and the opportunities existing for the construction of barrages for power generation across the rivers Severn and Mersey and elsewhere.

In the thirty years to 1975, RPT designed over 300 bridges for United Kingdom clients and a further 300 bridges were constructed overseas in more than thirty countries.

Northam Bridge, Southampton completed in 1954 at a cost of £500,000. The overall length is 485 feet. The deck is constructed of precast pretensioned concrete tee-beams connected together and post-tensioned transversely. The slender supports housed inside the piers are a development of Cuerel's Waterloo Bridge design permitting flexural movements.

Soon after the war the principle of using slender concrete supports housed inside the piers as at Cuerel's Waterloo Bridge, was taken a stage further at Southampton. Individual columns in bitumen-filled pockets were substituted for the continuous wall used at Waterloo. The 485 ft long Northam Bridge, which carries the main Southampton to Portsmouth road across the river Itchen, was commissioned by the County Borough of Southampton in the early stages of the town's post-war reconstruction programme, and completed in 1954 at a cost of £500,000. The bridge comprises three spans each 105 ft long and two side spans 85 ft long. The roadway has a dual carriageway and two footways giving an overall width of 66 ft. The deck is constructed using precast pretensioned concrete T-beams, connected together and post tensioned transversely. Prestressed insert slabs give continuity over the piers. The columns supporting the deck are housed inside the piers permitting flexural movements, and the piers are founded directly on a ballast stratum chemically consolidated as a protection against scour. Here, also, a new standard for concrete strengths was set. The water-cement ratio was required to be as low as possible, consistent with satisfactory levels of compaction and hydration and the concrete had to be literally dug out of the skips. The results achieved were remarkable, even by the present day standards, and were fully described by John Cuerel in his Presidential Address before the Reinforced Concrete Association. There would be much less deteriorated concrete everywhere if the standard insisted on by Cuerel had been universally adopted – short term economics have mitigated against it.

In the post-war years Ernest Bateson's department designed and supervised the construction of several important bridges in Iraq. Bateson was born in Bradford in 1895 and received his training at Bradford Technical College. He worked for Sir William Arrol for a number of years, and following a period in industry, he joined RPT in 1924. From 1946 until his retirement to his native Yorkshire in 1958 he was in charge of the Steel Bridge Department. The road bridges were of plate girder construction and the railway bridges had high-tensile steel trusses supported on concrete caissons. The Nasiriyah Road Bridge across the River Euphrates on the Basra to Baghdad highway was completed in 1959. It was an elegant structure having a central span 250 ft long and two anchor spans each 170 ft long. High-tensile steel was used for the plate girder work and the pile supported steel and concrete approach embankments gave an overall length of over 2000 ft.

Ernest Bateson's department designed several important bridges in Iraq.

In the 1960's, a similar bridge, known originally as Northgate Bridge, was designed to cross the River Tigris in the centre of Baghdad. Subsequently, the steel plate girder work was changed and replaced by six precast post-tensioned concrete beams per span connected together such that they became continuous under live load. Michael Morris who was intimately involved with the design and construction of the bridge is now employed in another company in the Group.

A period of seventeen years elapsed between the submission of the original report in 1957 and the final completion of the contract drawings and documents in 1974. This was by no means an abnormally long gestation period for such an important project in a Middle Eastern country. At various times the approach roads were realigned, extended, put on flyovers, widened to six lanes, reduced to four lanes, moved eastward away from the new hospital, moved westward away from the municipal buildings, and finally shortened so that the alignment reverted to that proposed in 1957. Even when ninety per cent of the contract drawings were completed, the whole bridge was moved 60 ft in a longitudinal direction because the river had been filled in on one side and the first one-and-a-half spans would have been on dry land.

The contract was finally let in November 1974 to a Joint Venture consisting of the Iraqi State Construction Company and a Japanese firm,

The 17th July Bridge, Baghdad carrying four carriageways over the River Tigris. Designed by Rendel Palmer & Tritton and completed in 1978.

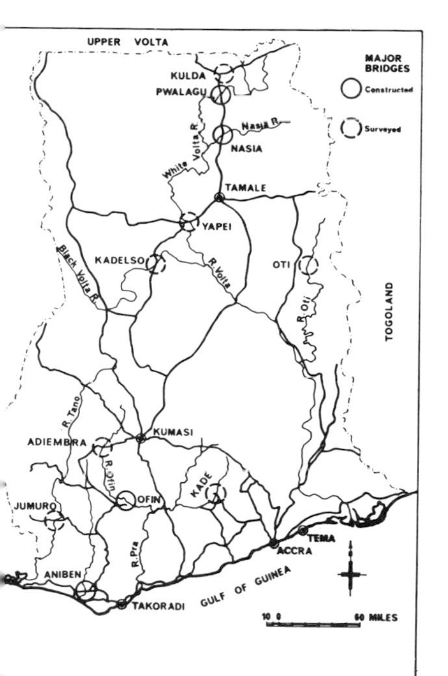

Location of major bridge works carried out by the firm in Ghana in the post war years.

Ohbayashi Gumi. Apart from some alterations to the design of the bridge foundations, the bridge was built in accordance with RPT's design, including the recommended method of construction. Shortly before the official opening in 1978, the Iraqi Government bestowed Northgate with a new name to commemorate the date on which they had assumed power, and henceforth the bridge was known as the 17th July Bridge. Downstream, its neighbour was named 14th July Bridge, recalling the earlier fall of the monarchy.

Paul Pilditch was for many years a key figure in the firm's links with the many client organisations in Iraq.

Other notable bridges designed by the firm were the 688 ft long Pazundaung steel cantilever road bridge in Burma completed in 1961, and the Sittang railway bridge completed in 1963. The latter, carrying the metre gauge railway on the Rangoon to Moulmein line, had six main spans of 360 ft 6 ins with high-tensile steel trusses fabricated in Yugoslavia. Roadways were cantilevered on each side of the bridge and the piers were constructed on wells with steel-cutting edges and strakes.

In the Gold Coast in West Africa, now renamed Ghana, the firm carried out surveys for the eleven major bridges shown on the accompaning map. Four bridges at Pwalagu, Nasia, Dunkwa and Aniben were constructed under the firm's supervision. A fifth bridge at Kade was designed by RPT and constructed under the supervision of the Ghana Public Works Department. John Palmer, F.P.'s eldest son, was responsible for the work. He was born in India in 1904, and at the age of 21, having obtained his Master of Arts degree in engineering, he joined RPT. Initially, he was assistant to one of the firm's inspectors working in Germany, where he acquired a first class knowledge of the language. He then spent two years in India working on the Gadavari Bridge for the Eastern Bengal Railway, before going on to Palestine, where he was employed on the construction of Haifa harbour and as a member of a survey team on the proposed Haifa to Baghdad railway project. Always a good mixer, these experiences were of immense value to him in later years when negotiating with overseas clients.

John Palmer – Partner 1946-1966.

John Palmer revered his father, but as a young man was completely overshadowed by him. Following his father's death he was advised to seek experience and broaden his horizons elsewhere. Almost brutally, Ernest Buckton made it clear to him that, as far as he was concerned, there was no scope for nepotism in the firm, nor any room for him in the partnership in the foreseeable future. Effectively, a similar thing happened to Julian Tritton for, as we have seen, the Senior Partner's job went to Ernest Buckton upon his father's death, and he in turn was succeeded by Ralph Strick. The agreement between the partners was that nobody should be admitted to the partnership without their unanimous approval, and that henceforth the Senior Partner was elected by the clear majority of his colleagues.

In 1937, John Palmer joined the staff of the Civil Engineer-in-Chief at the Admiralty and during the next nine years he held several important appointments, including that of Superintending Civil Engineer Thames, responsible for the construction of all works required by the Naval Staff in the Home Counties. His detailed diary was in due course to be the source of many fascinating articles he was to write for the RPT house magazine, *Rendels News.* Immediately after the cessation of hostilities in Europe, he was selected by the Admiralty to visit German naval dockyards, commercial shipyards and ports and to report on their development during the war.

Upon his return to RPT in 1946, his extensive administration experience, his knowledge of the design and construction of harbours, roads and

The Gongola River Bridge, Nigeria, carrying the 3 ft 6 ins. Gauge Bornu railway extension completed in 1962.

bridges, and his ability to organise large-scale surveys proved to be of enormous value to the partnership in a period of intense activity. Above all, he will be remembered as a man of great charm, with the priceless gift of dealing with people of all nationalities and all walks of life. In recognition of his many achievements in engineering he was awarded the C. B. E. in 1972. Although he served on the Council of the Institution of Civil Engineers for many years, circumstances dictated that he never achieved what would have been his great ambition of following his father as President.

Concurrently with the work in Ghana nine steel road bridges were designed and constructed in Sierra Leone. Plans were completed for similar bridges at Manowa, Mauwa, Mange, Sewa and Kumrabai.

Prior to the collapse of the Nigerian economy following tragic tribal conflict, and the world oil crisis, the Nigeria Railway Corporation employed RPT to carry out extensive surveys, prepare designs and supervise the construction of the 3 ft 6 ins gauge Bornu Railway Extension in Northern Nigeria on the Kafanchan to Maiduguri line. The most impressive structure on the line was the Gongola River Bridge comprising nine 150 ft long welded and riveted through truss spans, with floor systems connected by high-strength bolts, carried on concrete piers and abutments on well foundations taken down to the rock bed. A stone-pitched training embankment some 2000 ft long was constructed to guide the three-quarter-mile-wide river through the 1350 ft of bridge opening. Completed in 1962, the cost of the bridge was £800,000.

The extension started at Kuru, 4000 ft above sea level, and gradually fell to 1000 ft at Maiduguri, 370 miles to the north east. The work involved some 12 million cubic yards of excavation, about one-quarter in rock, and 7 million cubic yards of earthwork for embankments. Over seventy bridges with steel spans varying from 20 ft to 150 ft in length were built

The Kuru to Maiduguri line, Nigerian Railways, showing part of the 3 million cubic yards excavation through solid rock.

involving about 2200 tons of steelwork. There are also over 800 culverts, varying from 1 ft dia. pipe to multi-barralled reinforced concrete slab culverts up to 10 ft span. Station and service buildings, installations, and residential and office accommodation, in totality some 300,000 sq ft, were included in the project. RPT carried out the preliminary reconnaissance work, which was followed by aerial survey mapping and final field surveys. The work was divided into ten main contracts, nine of which were supervised by the firm at a cost of £10 million. The work was completed in 1964 and the Prime Minister of the Northern Federation, Alhaji Sir Abubakar Tafawa Balewa, boarded the first train from Kurh to Bauchi near his birthplace, a distance of 100 miles, in June 1961.

The railway's chief permanent way engineer, Hugh Alexander, now resident in Western Australia, recently recalled this journey, in which he was accompanied by Robert James, the railway General Manager, Ted Foster, the resident engineer, and John Palmer. He described the Prime Minister as a man of great dignity, having a depth of knowledge which few Europeans could match. They travelled in the General Manager's private coach, sitting on the veranda talking and acknowledging the cheers of the many thousands who lined the route. Upon arrival at their destination, John Palmer presented the Prime Minister with an engineering scholarship upon behalf of RPT, enabling a Nigerian student to attend a British university. (Foster later joined RPT and was for many years responsible for the firm's operations in Australia, Rendel & Partners).

John Palmer had a great knack of identifying with people.

John Palmer, in spite of his formidable style on first contact, had a great knack of identifying with people. He kept a card index of those whom he met so that on renewing an acquaintance he could talk about earlier personal circumstances. At opening ceremonies, in addition to an appropriate modest gift to the person performing it, he would often hand out silver penknives to senior personnel involved – but to avoid bad luck associated with the giving of a knife he would insist on the recipient paying him the smallest coin of the realm!

JP, always with an eye to the future, also used to try to ensure that young engineers, who could be important persons in their own country, in due course had an opportunity to be trained by RPT.

Many senior RPT staff were intimately involved with the extensive projects in West Africa, including de la Motte (killed in a plane crash whilst working there), Ronald Sweetapple (who was appointed a Partner in 1982 until his retirement in 1984) and Reginald Harvey. Henry Boyce assisted Palmer with many of his marine projects.

Before leaving overseas bridge work, mention must be made of two spectacular bridges completed in the Republic of Korea in the present decade. Both were funded by the World Bank as part of Trans Asia's work on the second Gwangju Regional Project. The Jindo Bridge, connecting the mainland to Jindo Island, with a main span of 1135 ft, has the longest cable stay span outside Europe and is the third longest in the world. The Dolsan Bridge, located near the fishing port of Yoesu, has a main span of 920 ft, an overall length of 1475 ft, and connects the island of Dolsan with the mainland. Both bridges are very narrow, being only two-lane, which makes them more difficult rather than easier to design. The box deck, towers and cables are similar on both bridges, but the main tower foundations are quite different. Advanced design pneumatic caissons were employed in the excavation of the main piers of the Dolsan Bridge, whereas the piers for Jindo are founded on land, out of the exceptionally strong tidal currents. RPT carried out the detailed design work and tender reviews for both bridges, undertook checks on the erection methods and supervised construction.

Mention must be made of two spectacular bridges completed in the Republic of Korea.

The Hyundai Construction Company were main contractors to the Jindo Bridge project. All the steelwork was fabricated at their huge Ulsan shipbuilding and heavy engineering works and brought to the site by barge. A huge 300 ton Manitowoc ringer crane with a 340 ft long boom was used for erecting the towers and sidespans. The erection of the superstructure was commenced in April 1983 and the completed bridge was formally opened by President Chun at a ceremony held on 18th October 1984. Within minutes of the President's departure a huge crowd of colourfully dressed spectators jammed the bridge providing a free, rolling, live-load test estimated at over 100% of the design live-load!

The contractor for the Dolsan Bridge was the Daelim Construction Company Ltd. In this case a 2000 ton floating crane was employed for

The Jindo Bridge, South Korea, the longest cable stay span outside Europe having a main span 1135 feet long. Completed in 1984.

The Dolsan cable stay bridge, South Korea. The erection of the superstructure was commenced in April 1983 and the bridge was formally opened by President Chun in October 1984.

erecting the towers and side spans. The tower base sections were the first units installed by the floating crane, and the fully pre-assembled towers were each erected as single lifts. Started in 1981 the work was completed in 1984.

During the design stages special attention was paid to the aerodynamic performance of the bridges, involving wind tunnel tests and extensive use of the firm's Prime 400 computer system for carrying out complex analysis. The result shown in the accompanying illustrations was an impressive, slender looking, aerodynamic box girder, supported at its edges by cables radiating from the triangular shaped steel towers, straddling the two-lane carriageways. Virtually all the design work for the two bridges was carried out in Korea under the direction of Richard Tappin who was later to be appointed a Director of RPT. Transferring new technology during both the design and the construction put heavy demands on Tappin and the RPT team, which they will long remember.

* * *

Bridge work carried out by the firm within the United Kingdom during the past thirty years falls into three main categories.

Firstly, they have been responsible for the design and detailing of more than 120 plate girder rail bridges for British Rail. The modernisation and electrification of the Eastern Region main line to the North and East Anglia has involved extensive bridge engineering. The majority of the work was necessary to carry the railway over or under existing roads, but in some notable instances the work coincided with the building of the M11 and M25 motorways and important new dual-carriageway by-passes at Royston in Hertfordshire and at Bury St Edmunds in Suffolk. With its wide experience of sliding bridges into position with the minimum disruption to rail traffic, RPT prepared detailed erection schemes for all projects and in most cases was retained to give additional technical support during construction.

The double track skew bridge built at Bury St Edmunds with a 154 ft long span, is one of the longest plate girder rail bridges in the United Kingdom. Similar bridges were designed by the firm for Colchester, Royston, St Neots and Waltham Cross. At the Old Ford Bridge on the new GLC East Cross Route, the substructure was installed horizontally by jacking through the embankment precast concrete box units which were subsequently filled with concrete and post tensioned. The superstructure consisted of thirty-six trapezoidal steel box girders bolted together to form a continuous deck system. The superstructure was constructed on each side of the railway, and each half of the bridge was then slid into position during a period of only forty-eight hours during a week-end track possession.

The nearby Victoria Park Bridge resulted in the firm being awarded a Constrado Structural Steel Design Award in 1973. It comprised a double track, seven span rail bridge, 525 ft long, across the East Cross Route on a very severe skew. The rectangular steel box main girders span 75ft and carry a prefabricated plated deck. The main girders are supported by transverse steel girders set on columns founded on bored piles.

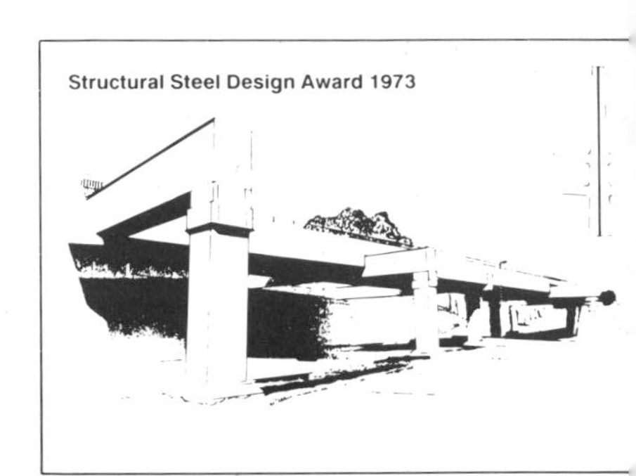

The Victoria Park Bridge resulted in the firm being awarded a Constrado Structural Steel Design Award.

Concurrently, the firm undertook the upgrading and recommissioning of a disused railway line to form a main route through the City of Belfast in Northern Ireland. This work included the six span 450 ft long Lagan Viaduct and four other underline bridges.

The second category of United Kingdom bridge engineering undertaken by the firm formed part of several major new trunk road schemes which the Ministry of Transport entrusted to RPT during the post war years. These included a twenty-five mile section of the 'Heads of the Valleys' A 465 Neath to Abergavenny trunk road; the A 48 Neath By-pass; and the A 470 Taff Vale trunk road running north from Cardiff through the valley of the River Taff.

The Taff Fechan Bridge on the Heads of the Valleys road was the largest of three similar structures in the contract, which in total included some fifty bridges. It had an overall length of 392 ft, and an arched span of 227 ft across the Fechan River. The twin parabolic arches with a 43 ft wide deck, were supported on portal frame spandrels, all in reinforced concrete.

RPT has been responsible for the design of 120 over and under bridges for British Rail Eastern Region during the past 25 years. The illustration shows the 47m single span skew Bridge at Bury St Edmunds, one of the longest plate girder bridges in the United Kingdom.

The arch ribs were cast in successive sections cantilevering out from the abutments. Each section was tied back by steel cables to a temporary tower, which in turn was anchored to the arch spring block. This ingenious and innovative method of construction resulted in not only an economic proposition, but an extremely elegant bridge, blending well with the surrounding countryside. The excellence of the design was acknowledged by a Civic Trust Award for Design.

The Nant Hir Bridge on the Heads of the Valleys road in South Wales. The structural design and elegance of this and two other similar bridges gained RPT the Civic Trust Award for Design.

Three bridges across the River Taff were included in the Taff Vale trunk road scheme in order to give access to communities living on the east side of the river. They had 180 ft long twin decks spanning the river and 135 ft long side spans. The structural solution of simply supported plate girder composite spans and the unusual twin-walled piers with the provision for jacking at their bases was chosen to cope with the predicted mining subsidence in the area of up to 10 ft.

The third category of bridge work undertaken included the structural survey of existing bridges and the preparation and management of schemes for the refurbishment or rebuilding of such bridges. Quantitative and qualitative assessments were carried out and the strength of the structure was then evaluated, determining permissible loadings and the estimated remaining life of the structure. In the 1970's six bridges constructed between 1768 and 1880 within the centre of Dublin were examined in association with N. O'Dwyer & Partners, and a report on their condition was prepared for the Dublin Port and Docks Board.

Two interesting United Kingdom assignments in this area were the examination and subsequent strengthening of suspension bridges across the Thames at Hammersmith and Marlow. The Hammersmith Bridge with a 400 ft main span and side spars of 141 ft was designed by Sir Joseph Bazalgette and completed in 1887. The timber deck is supported by wrought-iron cross girders and stiffening girders carried by hangers and mild steel suspension chains. RPT first reported on the condition of the structure in 1960 when weight restrictions were imposed. In 1970, following further deterioration, they advocated strengthening of the bridge. The G.L.C. called for a detailed scheme and work was commenced in 1974. Because it is a listed structure of architectural and historic interest,

RPT was responsible for the detailed scheme for the strengthening of Sir Joseph Bazalgette's Hammersmith Bridge originally built in the 1880's. Commenced in 1970 and completed in 1974, the bridge was kept open to traffic throughout the work.

great care was taken to preserve its appearance as closely as possible. Throughout the contract the bridge was kept open to traffic except for a limited number of overnight possessions. The stiffening girders were replaced with new girders giving greatly improved longitudinal stiffness. The timber deck was replaced by a hard-wearing surface similar to that used successfully on the Albert Bridge. Another major task was the replacement of the seized chain bearings on the tower tops.

The Marlow Bridge designed by Tierney Clarke had been completed in 1830. The superstructure required reconstruction because of severe corrosion of the anchorage steelwork and the failure of some of the suspender rods. New lattice stiffening girders were erected, and the existing cross-girders were replaced. A new steel battledeck was installed and rubber filled expansion joints were provided at the ends of each span. All the bearings, hangers and suspension chains were renewed, the chain tension being transferred to the rear of the anchor block by high tensile steel bars. Extensive reconstruction work was also carried out on the masonry towers. Again the work was done with the bridge remaining in almost continual use. There were many anxious moments when heavy vehicles ignored the special weight restrictions imposed, but fortunately no failures occurred.

It is interesting to recall that in the first volume of the Proceedings of the Institution of Civil Engineers, 1837-41, there is a record of a lively discussion on the stability of suspension bridges under cross winds. JMR strongly questioned some of the detail of the first Hammersmith Bridge. The topic continues to be of concern to designers and was a key factor in the design of the slender Korean bridges just described.

In Eire, the Corporation of Waterford commissioned RPT to survey and report on the Redmond Bridge, first opened to traffic in 1913. It consisted of a twin leaf opening flanked by six reinforced concrete spans across the River Suir. The steel and concrete structure was minutely examined and consideration was given also to the mechanical, electrical and operational aspects of the opening span. RPT recommended that prior to the replacement of the bridge, certain short term remedial works be carried out. Concurrently, the firm proceeded with the design of a replacement structure. A single leaf bascule flanked by seven precast

The reconstructed Marlow Suspension Bridge. RPT was responsible for the structural inspection and report, designs, drawings, contract documents and supervision of the reconstruction of the bridge.

prestressed concrete spans was chosen. The 82 ft long opening span was formed by plate girders supporting a steel orthotropic deck. New foundations using driven steel piles were included in the design, which was planned to be built in two stages, such that the existing bridge could be demolished upon completion of the first phase.

Before leaving bridges behind, we must return to the Indian Sub-Continent. During the tragic war between India and Pakistan, East Pakistan, now called Bangladesh, suffered severe disruption to its roads and railways, including the total or partial destruction of 276 bridges. In 1972, RPT was called in by the Overseas Development Administration to advise on the reconstruction of two severely damaged railway bridges which had originally been designed by the firm. The 5385 ft long 15 span through truss Hardinge Bridge across the Lower Ganges at Sara had one span down, and the similar 2350 ft long 7 span King George VI Bridge across the River Meghna, built in 1936, had two spans down. A team of RPR engineers was immediately sent out to Bangladesh and they quickly determined that as the monsoon season approached, it was of

The Hardinge Bridge over the River Ganges destroyed during the India-Pakistan War and rebuilt under RPT supervision in 1972. A total of 276 bridges were either destroyed or partially destroyed during the war.

The deepest caisson in the world under construction for carrying the 220 kV power line across the 7 miles wide River Jamuna in Pakistan, 1980.

vital importance that the broken spans were removed without delay. They feared that the obstructions would cause a scouring action around the pier foundations, which could result in further extensive damage. Contracts were organised and the work of removal was supervised.

Of the many RPT senior engineers who have worked in Bangladesh, Kenneth Cross established a special relationship with officials and people, and the firm's reputation there (and in many other countries as well) owes much to his ability to develop such enduring bonds.

The Merz Rendel Vatten (Pakistan) consortium, founded in 1948 and referred to at the beginning of this chapter, was short lived. Internal factions caused problems, but the main difficulties arose due to the economic situation in Pakistan. Although the final settlement of outstanding debts was long delayed, Merz and McLellan continued working in the country. In 1979, they commissioned RPT to review and update a caisson design produced some years earlier for the foundations of a 220 kV power line interconnector crossing the Jamuna River in Bangladesh. This resulted in the construction of the deepest caissons in the world, and will long remain an excellent example of RPT's remarkable innovative abilities.

The permanent banks of the Jamuna River are 7 miles apart at the site of the crossing, which comprised 10 spans of 4000 ft. The towers are 364 ft tall above the caissons.

The caisson foundations are required to carry the loads imposed by the towers and conductors, plus forces due to flow of the river. The foundation conditions are generally of fine sand and silt to a great depth, and this presented many problems. The foundations had to be designed taking into account scour at peak flood. The stability of the caissons resulted from the depth of embedment in the alluvial deposits below the scour level.

Reinforced concrete caissons of 40 ft outside diameter were finally selected, with a wall thickness varying from 7 ft 6 ins to 6 ft 0 ins at the top. The length of the caissons was 295 ft to 330 ft depending on the exact location. The caissons were sunk using a bentonite lubrication system in an annulus 3 inches wide round the whole structure apart from the bottom 10 feet. By this method the caisson sank under its own weight as material was excavated from the inside. Excavation was initially by grab, and at greater depths by a specially developed airlift. Once the founding level had been reached, the caisson shaft was plugged with gravel and the bentonite displaced by cement grout.

The interconnector which cost $67.5M enabled power generated from indigenous natural gas in the east to be transmitted to the western half of the country to replace power generated from imported coal and oil, thus saving precious foreign currency. Its cost was recovered in 19 months.

The technical success of the project was to lead to RPT being responsible for a bridge crossing the Jamuna River, as will be described later.

Ever since the days of JMR's involvement with railways in India and Sri Lanka and elsewhere in the 1850's, the surveying, construction and operation of railways has been a cornerstone of RPT's business. During the second half of the nineteenth century, due to the brilliant work of Sir Alexander Rendel, the design and construction of great bridges and the development of Indian Railways into a vast system having a route mileage in excess of 30,000 miles, made railways the mainstay of the business. Since then, with the advent and development of the internal combustion engine and the rapid development of civil aviation worldwide, the position of railways in relation to other forms of transport has become largely dependent upon government policy.

In three areas, the railway remains indisputably the preferred mode of transport, namely, for the bulk haulage of minerals and ores; for the fast movement of container traffic; and for rapid passenger traffic in densely populated urban areas. In some parts of the world heavy haulage freight trains, hauling in excess of 10,000 tons, are already commonplace. Large container ships carrying 3000 containers are daily circumnavigating the world, and the railway remains the ideal method of transporting containers to and from the ports. Technology in the twentieth century has to a great extent concentrated upon miniaturisation, the development of the micro-chip, the computer and the means of storage and transmission of information. However one of the greatest challenges in the twenty-first century will be the efficient, rapid transport of people in our great cities and urban conurbations. The delays and frustrations of moving people from their homes to their place of work is becoming increasingly intolerable, costly and damaging to the quality of life. The growth of greatly improved suburban rail services, light railways and metro type systems is both economically and socially essential. Even in developing countries like Nigeria, it reputedly takes up to five hours to cross from one side of Lagos, the capital, to the other. Already over 100 of the world's major cities have mass transit schemes in various stages of planning.

RPT has carried out numerous railway assignments worldwide.

Since the war, RPT has carried out numerous railway related assignments worldwide and their range of activities has included: feasibility and economic studies, the survey, design and supervision of the construction of railways; the reconstruction, upgrading and rehabilitation of existing railways and their infrastructure; and the modernisation of motive power, rolling stock and other equipment such as signalling and telecommunications. In this area the firm's inspection services have played a major role. Of equal importance has been the firm's involvement with the layout and equipping of container and freight terminals, loading and unloading installations for handling bulk materials and ore traffic, and the layout of sidings for ports, mines, power stations and other industrial users. A computer program has been developed to produce detailed print-outs of train running times, speeds, fuel consumption, horsepower and drawbar effort.

Mention has already been made of the important Bornu extension in Nigeria. Other post-war African railway projects have been undertaken in Ghana, Sierra Leone, Liberia and Kenya. In Ghana a 50 mile rail link from Achiasi to Kotoku, between Oda and Accra, was carried out by the firm. This shortened the railway between Takoradi and Accra by nearly 200 miles and involved the clearance of dense forests and heavy earthworks. The work was completed in 1956 at a cost of £3 million. Surveys were also carried out for the doubling of the track between Tarkwa and Manso, near Takoradi. The extension of the Awaso to Sunyani line, a distance of 115 miles, had been supervised by Francis Irwin-Childs 10 years earlier, when he was employed by the Ministry of Works.

In Sierra Leone, sections of the original 311 miles long Freetown to Pendembu line built in 1898 with steep gradients and sharp curves, was surveyed with the object of improving the alignment between Freetown and Bauya and at Tabe Bank, near Bo. The firm carried out site investigations and prepared contract drawings, but the construction work was carried out by the railway's own engineers. Subsequently, the Government took the decision to phase out the railway in favour of an extensive new highway system with which RPT were involved.

In neighbouring Liberia, RPT recently carried out a European Development Fund financed study into the feasibility of providing a

A 'Heavy-haul' iron ore train on the Lamco JV railway in Liberia.

110 km cross-country rail link between the existing Lamco JV railway and the Bong Mining Company's processing plant, thereby giving easier access to the country's capital and main port at Monrovia. RPT carried out ground surveys through thick bush using 'total station' instrumentation, and the contoured drawings and quantities were produced by computer.

★ ★ ★

During the 1970's, the Government of Kenya decided to develop the Kerio Valley, a part of the great Rift Valley, for industry and agriculture, and a new 240 km rail link to serve the area was considered essential. In association with the local consultants, East African Engineering Consultants, a feasibility study was completed in 1981. A feature of the line is the need for a 5 km long tunnel through the Kamasia Hills of volcanic deposits rising to about 500 metres above the general level of the land. The design of the tunnel has had to take into account the many geological faults, geothermal sites in the locality and the movement of hot liquors through the faults.

Railway construction began in Iraq in 1902 as part of the then projected Berlin-Istanbul-Baghdad Railway. Built originally to the metre gauge, the main line from Basra to Baghdad and Mosul in the north of the country was converted to the standard 4 ft 8½ ins gauge during the 1939-45 War. It became one of the main routes for Allied arms supplies to the Russians. After the war, the Iraqi Government formed the New Railways Implementation Authority to construct and put into operation new railways to augment the standard gauge and replace the remaining metre gauge networks. In 1951 RPT was commissioned to carry out a comprehensive examination of the system covering the economics of operations, finance and accounts, and development and modernisation, including motive power and rolling stock. The outcome was a plan to build 3000 km of new railway, and although the plan has been largely frustrated by the war between Iraq and Iran, the east-west lines from Kirkuk to Haditha, and from Baghdad to Husaiba on the Syrian border have been completed.

The outcome was a plan to build 3000 km of new railway . . .

Baghdad, the capital of Iraq, is a rapidly growing city of 3 million people. In 1980 it was decided that the first rapid transit system in the Middle East should be built as an important component of the comprehensive development of the city. In May 1981 a contract was signed appointing the British Metro Consultants Group (BMCG) as General Consultants. The consortium included W.S. Atkins & Partners, Freeman Fox & Partners, Sir William Halcrow & Partners, the Henderson Busby Partnership, Kennedy and Donkin, Merz and McLellan and RPT, with London Transport International and British Rail's Transmark as associates. The first stage of the system was to comprise two lines totalling 32 kms, some 38 stations and depots and workshops. The lines were to cross in the city centre and the north-south Thawia branch was to pass under the River Tigris. Had the war with Iran not intervened this would have been potentially a fine example of UK consulting engineers working in a major consortium promoting the image of UK Ltd in competition with groups from other nations which receive much more direct support from their home governments.

Concurrently, RPT and their Indonesian Associates, PT Indulexco, were retained to advise the Indonesian Government and to provide an operational analysis of all transport modes throughout the archipelago. Indonesian railways are mainly single track and comprise four separate systems all carrying mixed traffic. The Java network covers nearly 4000 route kilometres on which passenger traffic predominates, whereas the three networks in Sumatra are orientated towards bulk freight. All

In conjunction with a firm of Management Consultants RPT carried out a survey of the working of the existing line to the Port Dickson oil refineries in Malaysia which included a cadastral survey, traffic forecasts, engineering assessments and cost/benefit analysis.

traverse severe mountainous terrain. Studies of minor ports required RPT engineers to use the most basic means of transport to reach isolated areas not then having links to major land networks. Overnight accommodation was often primitive in the extreme. Stewart Pow of RPT led the team on this arduous study.

The 40 km long railway from Seremban, on the main line from Singapore to Kuala Lumpur, to Port Dickson on the west coast of Malaysia, was originally laid to sharp curves and heavy grades, using a light rail section and little ballast. Following the development of oil refineries at Port Dickson, RPT was commissioned to report on the working of the line, its present physical condition and its future potential.

In Western Australia three standard gauge heavy haulage railways operate for the transportation of the high grade iron ore found in the region of the Hamersley Mountains, to the ports of Dampier, Cape Lambert and Port Hedland. The 389 km long Hamersley Iron Ore Railway has its headquarters at Dampier, and the 426 km long Mount Newman Railroad and the 179 km long Goldsworthy Railway are based at Port Hedland. The considerable expansion of the Goldworthy company's operations in the past twenty years has necessitated the construction of a 67 km link from Mount Goldsworthy to the Nimmingarra Range, near Shay Gap. RPT was responsible for the route location, soils investigation, detailed designs of earthworks and structures and technical advice

A 12,000 tonnes payload iron ore train returning empties on the Mount Newman Railway, Western Australia hauled by a 3900 hp diesel-electric locomotive.

during construction. They also prepared designs for a 29 km extension of the line to serve deposits centred around Kennedy Gap.

For the Mount Newman railroad, RPT was also commissioned to report on problems experienced with the railway's 3900 h.p. diesel-electric locomotives and to advise regarding future purchases.

At home the firm was retained by the Northern Ireland Railways Company to report, design and supervise the re-construction of the disused Central Railway which traversed Belfast from east to west connecting the Belfast to Dublin and Belfast to Bangor lines.

In the 1970's British Rail announced its intention to rebuild the former Great Eastern Railway's London terminus at Liverpool Street and to combine it with the old adjacent L.N.W.R. Broad Street Station. The approach to Liverpool Street Station has for long been known to passengers and railwaymen alike as the 'Black Hole', and the new project necessitates increasing the number of lines entering the station from six to eight. The existing tracks covered by a multitude of bridges and passing through a deep cutting under the disused Bishopsgate Goods Depot, will all require realignment to give a straighter layout. This means that besides additional spans for some bridges being required, others will require complete reconstruction. RPT was called in to report on the feasibility of the bridge-works and to produce schemes for getting the two additional lines past the back of the existing tunnel, which will require the construction of a new curve between the old Broad Street line and the Bethnal Green to Cambridge line.

RPT was engaged in 1956 by the Southern Region of British Railways to report on a possible rail link between Heathrow Airport and Victoria Station. A report was prepared, showing the feasibility of such a link, laying additional tracks alongside existing lines from Feltham (where the tracks turned north-west to pass in the tunnel to the central complex of the airport) to Victoria, using one of the lines over the Grosvenor Bridge across the Thames and leading to new terminal platforms on the western side of the station adjacent to the airways terminal building then used by British Overseas Airways Corporation.

In the 1950s, there was a revival of interest in the Channel Tunnel.

In the 1950s, there was a revival of interest in the long history of the Channel Tunnel, and a Channel Tunnel Study Group was set up in 1957. This organisation commissioned a number of technical studies in which RPT, Sir W. Halcrow & Partners and Livesey & Henderson took part, together with French consulting firms. These investigations, into engineering design, geotechnical and soil mechanics problems, formed the main technical basis on which the present-day tunnel design is based. At a later stage, in 1970, British Railways had undertaken preliminary studies on a high-speed rail link between London and the Tunnel, and in 1973 RPT was asked to undertake surveys, preliminary designs, preparation of parliamentary plans and estimates for a 27 km section of the favoured route between Woldingham, 27 km from Victoria on the line to Oxted, and Tonbridge, on the existing main line to the Channel ports. The section included a 2.5 km tunnel through the North Downs and considerable changes to existing lines to make them suitable for operation of trains at 300 km/h.

Unfortunately for the firm the project was abandoned until it was revived in a different form by Prime Minister Margaret Thatcher in 1987. For the nation it was perhaps no bad thing that the earlier scheme was delayed because the overall concept lacked some of the necessary features evident in the later scheme.

At the time of writing, RPT in conjunction with another firm of consulting engineers, is preparing a scheme for one of the contractors bidding for the rail link between the Channel Tunnel and London.

* * *

During the past twenty-five years major road schemes have featured extensively in the firm's portfolio, with assignments in the Middle East predominating. Thousands of miles of formerly wild, inhospitable country, previously traversed only by camels and mules, often across mountains and desert, have been brought into use by the development of new highways to the great social and economic benefit of the local people. But briefly before considering overseas projects, mention must be made of important and prestigeous road schemes carried out for the Welsh Office.

In the 1960's the firm was responsible for the twenty-five-mile long section of the A 465 Neath to Abergavenny trunk road known as the Heads of the Valleys road. The carriageway is 33 ft wide in a formation width of 50 ft. The two-and-a-half mile section between Abergavenny and Brynmawr is formed on a shelf cut into the steep rock side of the Clydash gorge necessitating an extensive new embankment benched into the existing ground.

In the 1970 s RPT was commissioned to design and supervise the construction of the new A 470 Cardiff to Merthyr Tydfil trunk road, which included a connection to the Heads of the Valleys road at Merthyr. Kenneth Ainscow and subsequently Len Hinch, who was a Partner from

Len Hinch – Partner 1976-1984.

Part of the A465 Neath to Abergavenny 'Heads of the Valleys' road in South Wales.

1976 until his retirement in 1984, directed the sequence of contracts for the major project. The A 470 is a major radial route out of Cardiff and the new road by-passed the villages of Tongwynlais and Taff's Well bringing much needed relief to the villages during the rush hour periods. Historically the route of the new road between Abercynon and Merthyr is of interest, running parallel to the old railway opened in 1802, in order to relieve overcrowding on the Glamorganshire Canal, whose traffic was slow due to the large number of locks. Initially, the railway used horse-drawn trucks, but in February 1804 Richard Trevithick drove his high-pressure locomotive engine drawing a load of 10 tons of iron bars and seventy men for 10 miles between Pen-y-Darren Iron Works and Abercynon at an average speed of 5 mph. This was the first time a steam engine had been used to draw a load on rails. This milestone in our mechanical history won a wager of 1000 guineas for the owner of the iron works; poor Trevithick ending his days in a paupers grave.

The 22-mile-long dual-carriageway followed as closely as possible the line of the disused Glamorganshire Canal and some of the old railways in the valley, passing through densely wooded and uneven ground. Competition for space in the floor of the valley between the various forms of development and modes of transportation drove the new road part-way up the side of the valley in places, often where the worst geological conditions were encountered. The road was built to near-motorway standards with dual 24 ft wide carriageways. Altogether there are 7 interchanges, 36 bridges and 21 retaining walls having a total length of 10 miles. The road is illuminated throughout, using lighting columns and high masts. The work included the demolition of a seventy-year-old 100 ft high railway viaduct and the clearance of an old burial ground necessitating the re-interment of some 200 persons.

The road has been acclaimed as a superb example of engineering skill and imagination.

The completed road, sweeping through the spectacular steep-sided narrow valley has been acclaimed as a superb example of engineering skill and imagination. Its construction involved the stabilisation of the Taren Slip, some 8 million cubic metres of rock and soil which started moving over 6000 years ago. Sensitive public consultation was needed to allay the

A section of the A470 Cardiff to Merthyr Tydfil 'Taff Vale Trunk Road' in South Wales.

fears of the inhabitants of Aberfan that the extensive earthworks might cause a repetition of the terrible tragedy of 1966 when a coal tip overwhelmed the village school and its occupants.

Geotechnics played a major part in the engineering of this complex project. Geoffrey Walton, a Consulting Mining and Engineering Geologist was engaged to advise on mining subsidence. The Geotechnical Department of RPT was engaged throughout with the active involvement of Peter Fookes, Consultant Engineering Geologist who has served the practice since 1971. His eminence in engineering geology and the properties of soils and rocks is internationally recognised and there are very few of the firm's projects which have not benefitted from his wise advice.

In 1964, RPT was commissioned by the Chilean Government to prepare designs and tender documents for 560 km of new roads in the mountainous terrain between Temuco, 600 km south of the capital Santiago, and Puerto Montt. Beyond that point the only overland access to Punta Arenas, the most southerly town in Chile, a distance of 1350 km, was by means of a long detour through Argentina. The new roads ran

A night-time view, showing road lighting, of a section of the Taff Vale Trunk Road.

through the lush green valleys passing many attractive lakes and fast flowing rivers. They were always within sight of wisps of smoke from the many volcanoes in the area.

A general road study was carried out in Ethiopia in 1969, in the period before the Marxist Revolution, under the aegis of the United Nations Development Programme. The study included origin and destination surveys over the whole country, forecasts of traffic growth in relation to population changes and economic development, and an economic assessment and priority ranking of improvements in the transport system. The complete analysis of the system as a whole was one of the first to be carried out entirely by computer.

The following year, under the aegis of the International Bank for Reconstruction and Development (World Bank), a feasibility study and preliminary design for 1390 km of new roads in India was completed by the firm. This work was carried out by RPT's subsidiary company, Rendel Palmer & Tritton (India) Ltd., established in 1963. Richard Sarginson and later Colin Paterson were in charge in the Delhi office.

Three important feasibility studies for new highways in Iran were also undertaken by the firm in the 1970's. Following the annexation of British oil interests in Iran, relations between the two countries had been cool for some years, but as soon as the situation improved, RPT established a subsidiary company in Tehran, trading as Irendco. The first senior RPT man in Tehran was David Minch who later was to obtain a senior appointment in the World Bank. His successor was Ben Dixon-Smith who was Managing Director from 1969 to 1982 and on return to London Office was appointed a Partner. A feasibility study and detailed designs for the Kashan-Nain-Yazd road south of Tehran, in the direction of Kerman, was carried out, and the firm subsequently supervised the construction of 170 km of this desert road. In 1957, they also completed a feasibility study and traffic studies, together with some preliminary engineering work, for the 140 km of urban motorway around Tehran and 70 km of the Pazana major road across desert and wild mountainous terrain.

That roadwork was carried out by senior engineer John Warren who had the gift of being able to locate routes through difficult territory – mountains, deserts, jungle, coastal swamps. He was a quiet, unassuming man who would pass almost unknown in the office. He was happiest when, working with a small local team dependent on their own resources, he was meeting the latest technical challenge the firm had taken up.

In both India and Iran, recommendations were made as to the optimum balance between capital intensive and labour intensive methods of highway construction. These recommendations were derived from the analysis of the relative trade-offs between various standards of construction, and subsequent low-cost or expensive maintenance programmes.

In the 1970's the firm carried out three quite exotic road projects.

In the second half of the 1970's the firm carried out three quite exotic road construction projects. These were located in Nepal, high-up in the Himalayan foothills in the eastern part of the country; on the Pacific island of Fiji, and in Papua New Guinea.

Between 1976 and 1981, RPT engineers provided detailed drawings and technical advice during the construction of the $29 million U.K. Overseas Development Administration (O. D. A.) financed Dharan to Dhankuta road in Nepal, through severely mountainous terrain. This was probably the most comprehensive job of its size the firm had completely designed, drawn and documented almost entirely overseas. Only the proposed elegant 200 m long cable stayed Tamur River Bridge and some of the 16 km long Leoti River training works were designed in the London office. The Katmandu office, set-up by Kenneth Cross, employing locally

The Dharan to Dhankuta highway in Nepal high in the foothills of the Himalayan mountains in the eastern part of the country.

trained draughtsmen, produced over 500 contract drawings, which included 11 bridges and 400 culverts. Landslides, the icy green waters straight from the Himalyan snow melt, followed by monsoon rains and a thirty foot rise in the level of the river, turning it into a raging torrent, all became a part of everyday life for the engineers. The road and bridges survived very well in the disastrous 1988 earthquake which caused extensive damage, particularly in Dharan.

The World Bank financed the US $40 m Suva to Nadi Queen's Highway in Fiji, linking the capital and main seaport with the islands' international airport at Nadi and the intermediate coastal resorts, a total distance of 175 km. The two-lane single carriageway road traverses sand dunes, estuarine and mangrove swamps, and areas of deeply dissected hills requiring deep cuttings and large embankments of weathered residual soils. There are 65 bridges and over 1000 culverts along the route. RPT's contribution included an economic and technical feasibility study, topographical survey, geotechnical investigation and materials search, detailed design and contract documents for road, bridges and culverts, tender analysis and the supervision of construction.

Inspite of severe competition from no less than seventeen contenders, RPT's Australian office won the contract for a prestigious feasibility study for the construction of a new road linking Port Moresby with the major port and industrial area of Lae. The terms of reference required an initial survey of nine different corridors with recommendations for the most socially, politically and economically beneficial. The study team included sociologists, agronomists, environmentalists, political scientists, economists and engineers, illustrating the sophisticated and comprehensive nature of the services the firm was able to offer internationally. The survey covered an area of 50,000 sq km in the Owen Stanley mountain range between the Coral Sea and the Solomon Sea, several of the mountains rising to over 3000 metres. The area was almost entirely covered by dense forest and inhabited by a few tribes who still enjoyed inter-tribal fighting as a relaxation from their crude attempts at cultivation of the land.

Suva-Nadi Highway, Fiji: a 175 km two lane single carriageway traversing sand dunes, estuarine and mangrove swamps, and areas of deeply dissected hills.

Space precludes other than the mere mention of assignments connected with highway design and construction in Sudan, Syria, Iraq, Kuwait, the Yemen Arab Republic, Gambia, Malaysia, Indonesia and Hong Kong during the 1970's and 1980's. During this period the partnership's most important roadworks were in Libya.

In 1976 preliminary reports and designs were submitted for a 60 km length of road round the capital Tripoli and a 15 km urban ring-road at Benghazi. The Tripoli roads comprised the Second and Third ring roads each approximately 30 km long and designed to motorway standards. The Second ring road comprised 10 major interchanges, 30 bridges of both reinforced and prestressed concrete construction and two flyovers. The 30 month contract was let to the German contractors, Bilfinger and Berger, although a high proportion of the construction workers were recruited in Yugoslavia. Work on the Benghazi ring road started in 1980 and is still continuing. There are 9 bridges, 2 viaducts, and 7 interchanges. The contractor is a Libyan company working with a foreign contractor and maintaining this link has been a major cause of delay.

* * *

In the Hashemite Kingdom of Jordan, the name Rendel Palmer & Tritton is held in the highest esteem. Since 1960, the firm has been responsible for new road works, new railways, the development of the port of Aqaba, building the new airport at Aqaba and the establishment of bulk handling facilities for phosphate, the mainstay of the Jordanian economy.

In the 1950's Jordan was considered to be a country without a political or economic future. Indeed there were many who questioned the feasibility of its continued independence. It was in a state of war with its neighbours, especially the newly established state of Israel, it had an enormous and apparently insurmountable refugee problem, vast tracts of the country were desert, it had difficulty feeding its people, and it had no major industry or source of foreign currency of its own. It was entirely dependent upon both economic and military aid to survive.

Construction of some of the 1000 culverts on the Suva-Nadi Highway, Fiji.

The existence of phosphate and its value as an agricultural fertiliser was discovered by a Palestinean entrepreneur, Amin Kawar, in the early 1930's. During the harbour works at Haifa he developed a great friendship with Sir Frederick Palmer and this resulted in due course in the construction of a ropeway to the port of Haifa for the transportation of phosphate. After the war, with access to the Mediterranean denied them, Kawar's son, Tawfiq Kawar, energetically set about the development of new phosphate mines at El Hasa, mid-way between Amman and the town of Ma'an in the south of the country. Under the wise and courageous leadership of King Hussein, the state took an interest in phosphate and decided to develop the then little-used port of Aqaba at the head of the Gulf of Aqaba, having its outlet into the Red Sea. It is at this point Rendel Palmer & Tritton came into the picture. John Palmer took a great personal interest in these works and delighted in being called Abbu Aqaba – the Father of Aqaba. He was immensly proud when awarded the high honour of the Star of Jordan by the King. Michael Roberts, senior engineer, was Palmers' right-hand man for much of the firm's work in Jordan and whilst resident there was at one time honorary consul in Aqaba.

Tripoli second ring road, interchange No. 1.

The first stage of the development was the design and construction of the 200 km long desert highway across the Jordan desert from Amman to Ma'an. The road was completed in 1963 with two lanes. Apart from routine surfacing and the introduction of railway overpasses, the road has remained in service as originally engineered in spite of enormous increases in the volume of heavy traffic.

Benghazi third ring road, interchange No. 2.

The first ocean vessel cargo berth designed by RPT was opened at Aqaba in 1959, and in spite of the Six Days War with Israel, which brought shipping in the Gulf to a virtual halt, work went ahead with the construction of phosphate handling facilities, designed ultimately to handle 5 million tonnes per annum, with a ship loading rate of 2100 tonnes per hour. By 1984 phosphate exports had reached 3.15 million tonnes. In 1978, a potash berth was completed and is likely to develop into the country's second most important export commodity. Two slipways for small craft and ship repair facilities were opened in 1980, and the new container terminal and a ro-ro (roll-on, roll-off) berth were commissioned in 1984.

The first International Airport to be designed by RPT was opened at Aqaba in May 1972. Invited guests, including four representatives from the firm, were flown from Amman in a Royal Jordanian Airlines Boeing 707 with the King as co-pilot. At the airport, the King took the salute at a march-past before performing the opening ceremony, which included the award of the Star of Jordan to W.D. Goulden, RPT's resident engineer. After a never-to-be-forgotten traditional Arab luncheon the party flew back to Amman.

On the railway front the 270 km long 1055 mm gauge Aqaba Railway Corporation line from the El Hasa mine to Aqaba was opened in 1975. Initially seven trains ran over the system daily, comprising two General Electric diesel-electric locomotives hauling 30 to 34 forty-five tonnes payload wagons, substantially in excess of the gross weight of U.K. standard gauge freight trains. Very quickly the influx of valuable foreign exchange earned from the shipments began to reflect in the Jordanian economy.

The line from El Hasa to Ma'an utilized the old track bed of the Hedjaz Jordan railway, with a new extension to the mine and entirely new track from Ma'an to Aqaba. In 1980 the line was extended by 25 km to Manzil and two years later by a further 27 km to new mines at El Abyad. A feasibility study was also conducted with a view to extending the railway along the shore of the Gulf to the Saudi-Arabian border at Wadi Two. Currently, a great debate is underway as to whether the line should be continued as 1055 mm gauge, converted to 1435 standard gauge or made into a dual gauge line. There are also plans for refurbishing the exist-

RPT was responsible for the specifications and tender documen for the 18 diesel-electric locomoti and 210 covered hopper wagons by the Aqaba Railway Corporat for transporting phosphate from t mines at El Hasa to the Port of Aqaba, a distance of 270 km.

The 200 km long Desert Highway across the Jordanian desert connecting the capital Amman with the Port of Aqaba completed in 1963.

ing line north from El Abyad to Amman.

RPT has been concerned not only with the extensive civil engineering works associated with the railway, but also with the design and selection of locomotives and rolling stock and with the design and construction of the new locomotive workshop at Aqaba.

The firm was also engaged to prepare the designs for the wagon loading plant at El Abyad, incorporating a storage bunker and twin hoppers loading two wagons simultaneously. At Aqaba they designed for the railway an extensive phosphate handling system, including railway sidings, discharge hoppers, road vehicle tippers, an elaborate conveying system,

storage sheds, a phosphate stacking and reclaiming system and the shiploaders, particular attention being paid to dust suppression at every stage.

Nowhere is there a better example of RPT's professionalism and competence in a wide range of engineering and consulting disciplines than in Jordan. Events in this at one time 'country without a future' demonstrate just how much can be achieved for the benefit of man by good government, entrepreneural flair, the intervention of International Lending Agencies, and the professional competence of a firm like RPT.

Nowhere is there a better example of RPT's professionalism and competence.

Of great assistance to the firm's railway and highway engineers has been the introduction of the suite of alignment design computer programs MOSS, which could be combined with a computerised Digital Ground Model (DGM). Alternative routes can be imposed on the DGM and the print out of earth-work quantities aids the selection of optimum alignments. The selected alignment is automatically formatted and computer plotted on a contoured plan and the longitudinal section, with chainages and levels tabulated below. More recently the system has been extended to be interactive and IMOSS enables the design engineers to make changes on the VDU screen as their ideas develop and before printouts are made.

* * *

Throughout the history of JMR's and his successors' work, we have seen something of their wide and varied involvement in projects associated with the improvement of water resources. Originally, JMR's attention was directed towards the improvement of rivers and canals in the West Country; and in the Fenlands of East Anglia. In the twentieth century the firm's interests widened both geographically and in terms of the type of project undertaken and now encompass dam, irrigation, hydro-electric and flood control projects in a variety of environments.

Herbert Merrington – Partner 1958-1968.

In the 1960s, under the leadership of Partner Herbert Merrington, the firm carried out a major irrigation study and provided an assessment of possible future hydro-electric developments in Chile, near the ski resort of Portillo, some 160 km north of Santiago. The Chilean Copper Corporation's mines were located in the area, and at the time of the survey they were extracting 14,000 tonnes of copper ore a day and were anxious to find cheaper sources of electricity. Resulting from this opportunity an associate company was established in Santiago in 1965 trading as Rendel Palmer & Tritton (Chile) Ltd. Preliminary designs were prepared for development projects on the Diguillin and Nuble Rivers covering a total area in excess of 550,000 hectares.

Merrington, who was born in Haslemere in Surrey in 1904, had spent much of his professional career in India becoming Under Secretary of State to the Punjab Government, as Chief Engineer Irrigation. He joined the firm in 1949, became a Partner in 1958 and retired in 1968. It was not until 1978 that another Partner, Edward Haws, with special expertise in dams and irrigation was appointed. Both Merrington and Haws were appointed to the Senior Panel under the Reservoirs Acts.

Grouting trials being carried out for the Diguillin Dam in Chile.

Carrying on in the same tradition, RPT, during the last decade, designed, supervised and certified both the largest volume and highest UK water-retaining earth embankments, at Gale Common and Devil's Dingle respectively. Overseas, RPT carried out major flood studies for the Zambesi in Mozambique.

At the other end of the spectrum the firm undertook the design and inspection of remedial measures for small earthfill dams in the United Kingdom which required certification under current legislation.

In 1980 RPT formed with Parkman Consultants Ltd a joint venture to carry out irrigation and flood control projects; this was incorporated in

Artist's impression of the proposed 120 m high rockfill dam on the River Nuble, Chile.

1986. The joint venture undertook a broad range of research in connexion with a feasibility study, the detailed design and supervision of construction of the Arakundo–Jambu Aye Irrigation and Flood Control Project in northern Sumatra, Indonesia. The scheme will irrigate some 20,000 hectares.

In searching for water resources the team identified a hydro-electric site which was shown to have a potential to generate 160 MW.

At this stage RPT brought in Rendel Williamson Hydro Ltd, a company incorporated in 1986 from a joint venture formed in 1980 with James Williamson and Partners, to provide consultancy services related to hydro-power developments. The consultants put together, with contractors and a banker, a scheme for developing the hydro potential, and applied for Aid and Trade Provision finance from the UK Government; the expectations are that the project will proceed in due course.

The two joint venture companies have been responsible for other work in Indonesia, Chile, China and Pakistan.

Since 1984, Rendel-Parkman has been responsible for the co-ordination of a wide range of studies related to the river Mersey tidal power barrage project. These have included engineering economic, hydraulic, sedimentation, environmental, navigational and socio-industrial investigations. The scheme would generate between 480 MW and 750 MW depending on the final location and would make a major impact on the potential for economic and social regeneration of the Liverpool area. Second stage studies now in progress are to be completed in 1990. The project is likely to be the first tidal power barrage to be built in the UK.

Artist's impression of the proposed River Mersey tidal power barrage.

Throughout the world the rapid post war growth of civil aviation has

Jambu Aye main dam and power station, Indonesia.

been a source of many important commissions for the consulting engineer, and RPT has been in the forefront of developments. Work undertaken has included feasibility studies, air traffic forecasts, appraisal of aircraft characteristics, the identification of terminal and landside requirements, surface and tunnel access and environmental appraisal.

One of the earliest post war schemes undertaken was the development of Lungi Airport in Sierra Leone, West Africa. This was designed to bring the airport up to the international Class A standard so that the largest jet aircraft operating the transatlantic and similar routes could be handled. The runway was extended to 10,500 ft, the size of the apron was doubled and runway and taxiway lighting was installed.

In the United Kingdom, studies, and in some instances preliminary designs, have been undertaken at Gatwick, Stansted, Birmingham and at Maplin Sands in the Thames Estuary – the last being favoured by many as the ideal site for London's Third Airport before the final selection of Stansted. Overseas work has been undertaken in Cyprus, Gambia, at Juba airport in the Sudan and at Beirut, where a study for a possible seaward extension of the existing runway to meet the demands of larger aircraft and increased traffic was undertaken. Other studies were completed in the hostile terrain of the North Yemen, at Aden in South Yemen, in Iraq, Sri Lanka, Indonesia, Papua New Guinea, Fiji, the New Hebrides, the Gilbert Islands and in the Falkland Islands. In Iran, designs were completed and construction work supervised at a new hovercraft base.

The Falkland Islands project at Cape Pembroke Airport, near the principal town Port Stanley, was completed shortly before the Argenti-

Artist's impression of the firm's design for Amman International Airport, Jordan.

... an altogether remarkable achievement.

nian invasion. The new airport includes a 4100 ft long runway, an apron, a link taxiway and terminal buildings including a car park, road access and telecommunications and navigational facilities. RPT provided the original feasibility studies, surveys, engineering designs, estimates, preparation of contract documents, bid evaluations and supervision of construction – an altogether remarkable achievement when one considers the nature of the terrain, climatic conditions, the enormous distance from the United Kingdom, and the total lack of facilities then existing on the islands.

After the recapture of the islands from Argentina in 1982 the firm assisted the UK Government in deciding on the site of the new airport and reported on some aspects of the port facilities.

Lungi Airport, Sierra Leone.

9

Ports, Power Stations & Offshore Projects

We must now go back forty years in time to consider the part played by Rendel Palmer & Tritton in the development of Maritime Works, Power Stations and Offshore oil and gas installations, not forgetting the firm's Marine Department which for many years has played an important role in the design and development of specialised vessels. Started in Sir Alexander Rendel's day designing ferry boats and harbour craft, the Department today designs sophisticated vessels including those required to meet the needs of the vast offshore oil industry.

The three decades from 1950 and into the early 1980's were an extremely busy period for the firm's Docks & Harbours Department.

The three decades from 1950 and into the early 1980's were extremely busy for the Docks & Harbours Department.

The Tyne Dock deep water quay was designed to accommodate ore carriers up to 25,000 tons. It is 870 ft long and 94 ft wide and of open construction, having a reinforced concrete slab deck supported on vertical concrete piles 90 ft long. A curtain wall of prestressed concrete sheet piles was built into the deck slab at the rear, the whole being tied back by pre-tensioned high-tensile steel cables to concrete anchorages. The firm carried out the site investigations, design work and concrete drawings and supervised the construction of the quay and bunker foundations.

At Port Talbot in the Severn Estuary, South Wales, a new ore terminal was designed to accommodate vessels up to 60,000 tons. The original entrance lock through which all iron ore for the steel works passed limited the size of bulk carriers to 10,000 tons. RPT's proposals provided a tidal harbour with breakwaters taken to the 3 fathom mark with an approach channel dredged 26 ft below datum, and an ore discharging berth with a maintained dredged depth of 44 ft. In addition, the plan provided for an area of water within the harbour which could be developed for further deep water berths if required. Wave penetration and sand movement in the harbour which were much debated at the time have proved to be very close to the RPT predictions.

Henry Boyce and Derek Kirby-Turner, senior engineers, assisted John Palmer in this project and in developing and controlling many of the port projects for which he was responsible. (Kirby-Turner collapsed and died during a visit to the Limassol Port Project, Cyprus, for which RPT was responsible, completed in 1973.)

In India, the firm designed an ore loading plant for the Madras Port

Tyne Ore Quay — a deep water quay for discharging iron ore, consisting of a reinforced concrete deck supported on piles.

Trust suitable for exporting 4 million tons of ore a year. The ore was to be delivered by rail wagons to stockpiles from which it was reclaimed for loading.

Since the earliest days of coal fired steam ships, Aden has been an important refuelling port of call. In 1954 the firm completed a new oil harbour for the then Anglo-Iranian Oil Company. The harbour provides berths for four 32,000 ton tankers and is protected by a 4126 ft long rubble breakwater. The jetties, supported on hexagonal piles, are of all-welded steel framed construction with concrete decks. Steel fender piles driven to a batter in front of the berthing faces transmitted the berthing force to the main framework through compression rubber blocks 32 ins diameter by 32 ins long. In 1981 the firm was called in to prepare schemes to rehabilitate the jetties to deal with the ravages of time and use in the harsh climate. These works are to be completed in 1989.

The hexagonal piles were a good example of the old adage that 'Necessity is the Mother of Invention'. Originally, it was intended to use tubular piles for the jetties, but as a result of the shortage of suitable capacity in the post-war steel industry, they were not available in the United Kingdom. John Cuerel and John Palmer had meetings with the South Durham Steel Company and persuaded them to roll a special section which could be welded together to form a hexagonal pile measuring 12 ins across the flats. Although RPT declined to patent the idea or to levy any form of royalty for its use, so that clients could not accuse the firm of benefiting by specifying their own design, the name 'Rendhex' pile was established and it became widely used.

Improvements to bumper fenders and dolphins, designed originally by Ernest Buckton during the work at Haifa in the period 1935-37, were also actively under consideration at this time. However, it was considered unwise to proceed with any new patent application, and this RPT innovation also became widely used worldwide.

Port Talbot Harbour, South Wales. The project involves a total length of 2500 metres of rubble breakwater which forms a harbour containing a jetty for unloading ore carriers. The ore berth is reached by a 215 m long approach jetty.

Other major port and harbour works completed in the 1950's and 1960's were at Takoradi in Ghana (then still the Gold Coast) and for the booming oil industry, which was expanding its loading and unloading facilities worldwide. In view of the fact that in 1969 the firm was designing a jetty at Finnart to accommodate a 500,000 dwt tanker it is fascinating to note that a 1950's letter from British Petroleum stated that they did not contemplate tanker sizes exceeding 80,000 dwt!

At home important developments were completed for the commercial ports at Swansea, Leith and Aberdeen. Major schemes for the bulk handling of iron ore imports were also well advanced. These included the Red-

Isle of Grain Refinery Jetties, Kent — seven deepwater berths for tankers up to 45,000 dwt in 12 m depth of water. Included in the project, completed in 1959, were the foundations of the entire refinery.

car terminal for supplying the Teesside steel works, and at Hunterston for supplying the giant Ravenscraig complex near Motherwell in Scotland.

At Takoradi the works, completed in 1954, included an additional 1400 ft long deep water quay to provide three more important berths, two of them equipped with double-storey steel-framed transit sheds, and five shallow water quays each 500 ft long for timber exports. Steel framed

Aden Oil Harbour which provides four steel piled berths for oil tankers and is protected from the monsoon storms by a 4000 foot main breakwater.

Khor-al-Amaya Oil Loading Terminal, Iraq. One of the major loading terminals off the coast of Iraq, Khor-al-Amaya is an island terminal 28 km offshore in 23 m of water.

sheds were also provided for the storage of sawn timber which, together with coconuts, had become the country's main exports. One of the double-storeyed transit sheds was equipped on its upper floor for passenger handling and disembarkation. Other facilities included the extension of a berth for oil tankers and a bauxite loading berth. The firm was responsible also for the extension of the port's railway facilities and new locomotive sheds. This work followed on from the original construction of the port in 1928 for which RPT had been responsible and was probably one of the last major projects the firm carried out under the aegis of the Crown Agents in a, then, British colony.

Notable work for the oil companies included the North Pier at Mina-al-Ahmadi in Kuwait. This included berthing accommodation for four 100,000 ton tankers or five 65,000 ton tankers in 60 ft of water. The pier, constructed of steel piles with a welded steel superstructure, has an approach arm nearly a mile long. The project included electric light and power installations, cathodic protection, gravity fenders and specially designed counterbalanced loading booms for high loading rates.

Between 1950 and 1958, six T-head jetties were completed at Bandar Mashur in Iran. Nos. 1-4 jetties, constructed on steel box piles filled with reinforced concrete, provide berthing for 30,000 ton tankers. Nos. 5 and 6 jetties have upper and lower decks and are suitable for 45,000 ton vessels. Main breasting dolphins are provided at the ends of the T-heads and gravity fenders, which had been developed after the war, are provided in the berthing faces of the earlier berths whereas the later two have steel frame fenders backed by rubber blocks, which are now almost universally used for fendering systems. Each jetty has Woodfield hose handling equipment for loading the tankers.

RPT also advised on possible sites for oil terminals in Iraq. A site at Khor-al-Amaya was finally selected. Situated 20 miles south of Fao in the open waters of the Persian Gulf, the project involved problems of strong tidal currents and the rate of silting from material brought down by the Tigris and Euphrates. Another site, Khor-al-Khafka, which has since been

Marine Terminal, Angle Bay, Milford Haven — two berths for 100,000 dwt tankers in 16 m of water.

used for very large tankers of 300,000 tons, was not adopted in the first stages of the oil export boom because this would have involved a much longer underwater pipeline from Fao. It became the first occasion in which De Long pontoons on jack-up platforms were used for an isolated offshore terminal. They were named after Colonel De Long who invented synchronised climbing jacks so that when in position the legs are dropped and the pontoon climbs up the legs. They had previously been used for jetty heads by the U.S. Army in Greenland and for an ore loading terminal in Venezuela, but the approach to the jetty heads was short. The pontoons were fabricated in the UK, loaded with other construction materials and plant and towed to the Gulf. They served as temporary working platforms before incorporation in the finished works.

At home, the oil terminal work completed included five jetties at Swansea, replacing old timber jetties with reinforced concrete structures. The piles were spun reinforced concrete tubes filled with bitumen after driving. A £2 million jetty was completed at Angle Bay, Milford Haven, in 1960, able to accommodate two 100,000 ton tankers. The firm was not only responsible for the initial land and marine surveys, the jetty design and supervision of construction, but also the preparation of a Private Member's Bill – ably demonstrating that every aspect of a complex project was within their competence. Other important oil terminal projects were completed for British Petroleum at Finnart on Loch Long in Scotland, at the Kent Oil Refinery and at Kwinana, and Westernport in Australia.

The handling of general cargo has undergone significant change since the War. Containerization and roll-on/roll-off operations have greatly increased cargo handling rates which has resulted in their adoption worldwide. This, in turn, has led to big increases in the size of container vessels, some of which are now capable of carrying in excess of 3000 containers. In the 1960's, with over 100 years experience in port and harbour work behind them, RPT became one of the leading consultancies in the world able to deal with port development and the modernisation of the landside operations.

RPT became one of the leading consultancies in the world able to deal with port development.

Royal Seaforth Dock, Liverpool — an impounded basin providing ten specialised berths for container and ro-ro traffic, bulk grain and frozen meat, packaged timber and general cargo.

Much of the cargo handling was directed by Kenneth Dally who first joined RPT in 1961. From 1967 to 1981 he was Assistant Director, Technical Services, National Ports Council, responsible for the cargo handling and port operations. He rejoined the firm in 1984 and was a Partner until his retirement in 1988. Charles Howard, a senior engineer in the Department, was responsible for many of the Materials Handling projects which the firm carried out. As noted earlier he was killed in a hotel fire in South Korea whilst inspecting steel fabrication there.

In the 1960's our West Coast ports were still of great importance to our Trans-Atlantic trade. Prior to his retirement John Palmer had made proposals to both the Mersey Docks & Harbour Company and the Port of Bristol Authority for major new docks to deal with large modern container vessels and bulk carriers. He provided tentative layouts of the basins and preliminary estimates of cost, including a rate of £750 per foot for the quay walls. He had obviously had in mind conventional methods of construction, with steel piling for the walls tied back to dead man anchors, the basins being excavated by dredger in the wet.

At Liverpool, the client was anxious to avoid the use of anchor bars behind the walls which might inhibit or complicate the foundations of building or other structures. It was realised that there could be difficulties with the steel piling having to penetrate hard clay, then cobbles before driving into sandstone, and furthermore there was a noise embargo on pile driving by percussion. A large volume of material was required to make up the areas behind the quays, and if the excavation had been done in the wet, most of the clay would have been useless for that purpose. The quay walls therefore had to be self-stable and able to penetrate through clay boulders into sandstone and be able to resist external water and earth pressure without leakage if the excavations were to be done in the dry.

This was the situation inherited by Francis Irwin-Childs when he took over the project from John Palmer in 1966. He produced designs for a quay wall having an arched profile with a rear fin when viewed in plan, to be sunk from the surface by in-situ diaphragm wall construction. The shape of the wall soon became christened the 'the wine glass'. It did not need long anchor ties extending back under the quay where they would obstruct other construction works. With the diaphragm wall process trenches are excavated in the ground and stabilised during the operation by a thixotropic suspension of mud, normally bentonite. Concrete is then inserted into the bottom of the trench displacing the slurry as it rises to leave a panel of concrete embedded in the ground. Such panels can be taken to a considerable depth. This construction technique can, in appropriate locations, show considerable cost savings in the construction of marine works compared with more conventional methods.

This construction technique can, in appropriate locations, show considerable cost savings.

At that time most applications of diaphragm walling used it merely in plane form as an improved type of sheet piling which had to be similarly tied back with horizontal ground anchors. In Irwin-Childs' 'wine glass' design, with the panels turned normal to the wall face as buttresses, they served as anchors to the structure, dispensing with the ties, because it is virtually impossible to rotate a panel through the ground along its major axis without shearing the soil in either side.

The Contractor and the Harbour Engineer were concerned about the stability of the proposed design and Harold Scrutton, only recently appointed Senior Partner of RPT, was informed that if he allowed things to go ahead, he would personally be held responsible for the consequences! Quoting Irwin-Childs' own words, 'I was summoned into the presence to be told that the Senior Partner was a conservative man and it would be a good idea to forget this unproven idea.' Whereupon Irwin-Childs obtained approval for a trial panel to be cast and test loaded. It was entirely satisfactory and the project proceeded, producing a saving in cost of £2.5 million, the cost of the quay walls being about £500 per foot.

Other unique features of this project are the 130 ft wide sector gates in the passage from Gladstone lock. They were designed by Brian Holloway, the partner responsible for steel structures, after an inspection of similar gates in the St Lawrence Seaway in Canada, and are believed to be the largest in the world.

The result of this work was the Royal Seaforth Dock, completed in 1973. It provides ten specialised berths for container and ro-ro traffic, bulk grain and frozen meat, packaged timber and general cargo, with a capability for further expansion. The impounded basin is enclosed by a 3 km long rubble mound breakwater. The firm's services included the planning, design and construction supervision of the docks and the huge automatic impounding pumps, quayside container cranes and extensive electrical and mains services installations.

The title 'Royal' was bestowed upon the docks when they were graciously opened by HRH the Princess Anne on 18th July 1973. Subsequently, Francis Irwin-Childs and M. Agar of the Mersey Docks & Harour Company were awarded the Institution of Civil Engineers Robert Alfred Carr Premium for thier paper 'Seaforth Dock, Liverpool – Planning & Design, and Brian Holloway and his assistants John Wadsworth and Michael Feit were awarded the Institution's Halcrow Premium for their paper 'Passage Gates for Seaforth Dock'.

Field-Marshall Sir Gerald Templar KG, HM Lord Lieutenant of Greater London, presenting the Queen's Award to Industry, 1973, to Mr. Francis Irwin-Childs, Senior Partner.

1973 was also an important milestone in the firm's history, for that year they were awarded the Queen's Award to Industry for Export Achievement. The Award was presented to the Senior Partner by HM Lord Lieutenant of Greater London, Field-Marshal Sir Gerald Templar, K.G., at a reception for the firm's staff and their partners held in the Great Hall of the Institution of Civil Engineers on the very site of JMR's home during

Princess Anne unveils (on July 18th, 1973) the title plaque which was eventually mounted at the main entrance to Royal Seaforth Dock.

his early years in London. In making the presentation upon behalf of the Queen, Sir Gerald referred to the firm's 'enormously distinguished record both at home and overseas, dating from 1838, the Coronation Year of Good Queen Victoria, and in particular to the recent efforts which had doubled the volume of its overseas work between 1970 and 1972, inspite of increasing overseas competition'.

The second project Irwin-Childs inherited from John Palmer was the new dock and lock at Portbury for the Port of Bristol Authority, to be built on marshland at the confluence of the rivers Severn and Avon. A spring tide range at the rate of 13.2 m, one of the largest in the world, provided a major challenge for design as well as construction. Palmer had originally put forward a large scheme for two basins and two locks which was rejected by the Ministry as being too extravagant. The Authority then advised RPT that they would be seeking advice from another consultant. Later, with Seaforth well under way, the firm enquired about the position at Portbury and were informed that the new consultant working with a major contractor had put forward a proposal for a single basin with a single lock at half the price of John Palmer's scheme and that this too had been rejected. It now became clear that, if the scheme were to go ahead, the current rates for quays and locks would have to be virtually halved.

In 1971, the issue was then taken to the House of Lords, where it was bitterly opposed by other port authorities in South Wales. The case hinged on their claim that the money should be spent on them rather than the Bristol Authority. The answer given so satisfied their Lordships that approval was given without waiting to hear the closing speech of Bristol's counsel.

The project proceeded using diaphragm wall techniques resulting in a dock covering 28 hectares having six berths each 308 m long served by the largest lock in the United Kingdom 365 m long by 42 m wide by 20 m deep and suitable for vessels up to 75,000 tons. The 4 million cubic metres

Redcar Ore Terminal – Berth for 220,000 dwt vessels, 320 m long berth in 23 m of water. Believed to be the deepest solid type quay using diaphragm wall techniques.

of dry excavation removed from the dock were used for the reclamation of over 220 hectares. The project included unique steel mitre gates, a pumping station, sluices, and a tidal pond. The purpose of the pond was to retain water at high tide level and allow the silt to settle out before the water was fed to the basin and the lock to maintain impounded levels.

On 8th August 1977 the new dock was graciously opened by HM the Queen who arrived in the Royal Yacht *Britannia.*

Until very recently traffic through the dock was light. Now following completion of the M5 motorway to which the port has direct access, industralists are building on the reclaimed land and the benefits of the project are being reaped.

The project was awarded the rare 'Snow Award for Innovation' by the Concrete Society in 1978, the judges recording their 'delight' at the concept. In presenting the Silver Goblet the Duke of Gloucester wondered, 'what on earth the consultants would do with the vast amount of money that had been saved? The Senior Partner's response was that, as an architect, HRH would know that sadly the professional advisers don't get it!'

Consequent upon the success of Seaforth Dock, RPT was invited by British Steel to design an ore terminal at Redcar on the River Tees to accommodate 220,000 ton ore carriers. Existing berths for vessels of this size had been built a mile or more from the shore, with piled berthing heads for heavy cranes. Trestles with multiple conveyors delivered the ore to stock piles at the works. The Chief Engineer was heard to wish that it were possible to dispense with the trestles and bring the huge vessels in shore, and what an enormous advantage that could be. The initial requirement was to determine the feasibility of construction and maintaining a

Hunterston Ore & Coal Terminal Jetty – a deepwater terminal in the Firth of Clyde; one of the largest structures of its kind in the UK.

stable channel from the deep water to the shore. RPT researched the history of the Tees since the town of Middlesborough was purchased by the Quaker Pease family in 1829 in order to establish a new coal exporting port. Many of the great names in civil engineering had been involved in the intervening years, including JMR, Sir Frederick Palmer and James Buckton, in transforming a treacherous waste of shifting sandbanks to the present day controlled deep channel, and it was eventually concluded that a new stable channel could be established for the giant ore carriers. For the berthing structures, the pocket would need to be some 30 m deep from the cope, and the crane loads on the front rail would be high because of the outreach to the large carriers. Although cellular walls of steel sheet piling have been built, they were far too shallow for this application. The steel works were asked with some trepidation how they would feel if the concept was developed in concrete. The reply came that their main concern was a satisfactory structure.

The concrete panels of the cellular walls are 43 m deep and the structure is still believed to be the deepest solid berth in the world. The project was also notable in that the estimated cost in July 1970 was given as £3.440 million. The final cost four years later worked out at £3.438 million.

In the same year RPT was commissioned by the Clyde Port Authority to design and supervise the construction of the marine section of the Hunterston ore terminal with facilities for berthing vessels of up to 350,000 tons. The jetty head has been sited to provide a minimum of 40 metres of water at its outer berthing face, and lies 1½km offshore and handles both incoming and outgoing cargoes. The 443 m long by 34 m wide jetty head is of precast concrete construction on large diameter tubular steel piles. The total cost of the project, which was opened by the Queen Mother in 1979, was about £30 million. The RPT resident engineer was Neil Robertson, whose last project it was before retirement, having carried out a similar role for the firm at Westernport and in Iran.

The project was opened by the Queen Mother in 1979.

During the 1960's RPT carried out extensive works for the development of the shipyard of Cammell Laird at Birkenhead to provide facilities for the construction and servicing of Polaris submarines and the larger bulk carriers then being developed throughout the world.

The Princess Dry Dock constructed for Cammell Laird & Co. at Birkenhead. Length 950 feet, width of entrance 140 feet.

A tanker cleaning berth was built on two 200 ton strongpoints, spaceframe structures prefabricated and lifted into place by a floating crane. The design of the spaceframes required many days of tedious hand calculations – nowadays the work would be done by computer in as many hours and with more sophisticated parameters.

Two construction slipways were reconstructed for the building of the submarines and 100,000 dwt bulk carriers. Structures on large diameter bored piles carried 100 ton travelling cranes to handle prefabricated units constructed in adjacent workshops instead of the vessels being formed from piece-small units on the slipways. Technology has continued to advance rapidly since those structures were built and the unit sizes which were considered large at the time would now seem trival. Submarines and many vessels are now built totally undercover.

The Princess Dry Dock, so named following the opening ceremony by Princess Alexandra, was built on the site of an earlier dry dock. It is 290 m long by 45 m wide. A feature is a 1700 ton banana-shaped steel sliding entrance gate of some 60° arc length designed to avoid interference with and adjacent dry dock.

The firm designed and supervised the construction of all the works at Birkenhead including the complicated power and piped services, cranes and pumps.

The same boom in ship construction and repair saw RPT designing and supervising for Harland and Wolff, the construction of the 335 m long by 50 m wide East Twin Dry Dock at Belfast to accommodate tankers of up to 200,000 dwt. The Cammell Laird dock structure had been relatively simple to design as it was largely hewn in red sandstone. The structure at Belfast was very different as it was in softer ground. Much of the floor had to be anchored down to resist the uplift water pressure and the steel piled walls needed sophisticated anchorages to resist the pressure from the local alluvial material known as 'sleech'.

Sadly, the boom time for British shipyards has passed and both those yards are now shadows of their former selves.

Many ports in the United Kingdom were developed in the 1960's to accommodate the new generation of cargo vessels which had a draft of about 13 m. RPT had a major share of that activity as is evident from other projects described below.

East Twin Dry Dock, Belfast. At 335 m long with a 50 m entrance width and 11.6 m of water over the sill at spring tides, this is the largest repair dock in the United Kingdom.

Fowey Port Development, Cornwall – development of a china clay handling facility, involving 168 m of piled quay, re-decking of a jetty, construction of a clay store and installation of handling equipment.

In developing Leith Harbour the Forth Ports Authority had the choice of opening up their tidal docks (as was to be done at Aberdeen for lesser draft vessels) or constructing a new lock entrance to impound a much larger area within the breakwaters which had been built during the war. The RPT report on the project included what was probably the first detailed economic study the firm had been required to carry out for such a scheme. It served its purpose but in retrospect it would not meet the more rigorous requirements which such studies now have to satisfy. The new lock entrance was the preferred scheme. The lock 260 m long by 33.5 m wide was constructed inside a cellular cofferdam proposed by the contractor Edmund Nuttall. Significant economies were achieved by later using the cofferdam material in place of the original design to close off the remainder of the entrance to effect the impounding of the harbour. The lead-in jetty was formed from large diameter high tensile flexible steel tubes to provide the necessary energy absorbing fendering. This system had been developed by Mannesman in Germany but the tubes for Leith (and jetties at West Thurrock for which RPT was also responsible) were made by the British Steel Corporation then developing its techniques.

Only brief reference is possible to other interesting port and harbour projects undertaken by the firm in the 1970's. At Fowey in Cornwall the china clay handling facility, built originally by the Great Western Railway during the First War, was modernised. 168 m of new piled quay was constructed, the jetty was redecked, new conveyors and a shiploader were built, together with a new clay store and other modern handling equipment. Vessels can now be loaded at a rate of 1000 tons per hour with the wide variety of bulk and bagged clays which industry requires.

Leith Harbour was developed as an impounded basin with a 300 m long cellular sealing dam and a new 33.5 m wide entrance lock. The photograph shows the lock during construction.

For the same client, English China Clays, RPT, when asked to report on the development of Par Harbour, was able to show the manuscript report which the firm had prepared last century concerning the construction of the harbour to suit small coastal sailing ships!

On the other side of the world, in Australia, the Hay Point Coal Terminal for the Utah Mining Co. Ltd. was designed by RPT engineers. This deepwater terminal for the export of 20 million tons of coal per annum has two berths for 100,000 ton and 120,000 ton bulk carriers in 17 metres of water with a 1830 m long approach arm. The first berth was constructed using conventional steel pipes but in order to avoid problems encountered RPT designed the second berth to consist of prestressed concrete caissons, built in MacKay Harbour, and towed the 24 km to the site and sunk on to a prepared bed. The caissons are designed to withstand cyclonic waves 7.5 m high. The mechanical handling equipment includes two shiploaders of 5000 tons per hour capacity and conveyors from the shore to the berths. Much of the equipment for the berths themselves was erected on the caissons before they were towed out, thus avoiding expensive offshore erection.

In India the firm completed the first stages of the new dock complex at Haldia situated at the junction of the Rivers Hooghly and Haldia, 55 miles downstream from Calcutta. The function of this new dock system is to allow the berthing of large modern bulk carriers and tankers unable to reach the Port of Calcutta, and to overcome the problem of progressive silting in the upper reaches of the Hooghly, towards Calcutta. The first stage of the impounded dock provides six berths for coal, ore and phosphates, as well as general cargo, all with a constant water depth of 13.7 m. The entrance lock has a length of 300 m and a width of 39.5 m, with three sliding caisson gates. Allowance has been made for a considerable increase in the berthage during later stages together with a dry dock and a second entrance lock.

Also in India, starting in 1965, RPT was responsible for the development of a completely new port at Paradeep some 300 km south-west of Calcutta. This site comprised creeks, coastal marshes and sand dunes. The port was primarily for the purpose of exporting iron ore but the comprehensive development scheme provided for 19 cargo and 1 oil berth. When work commenced financial stringency determined that the initial excavations for the basins down water level should be carried out by local labour and at the peak some 10,000 coolies with primitive tools and headpans were employed. Much of the rock for the breakwaters, one of which was 1800 m long, was manhandled after it had been off-loaded from lorries which had brought it tens of miles from a quarry. The west coast of India has a massive littoral drift from south to north. Initially RPT designed an offshore breakwater to form a trap from which sand could be dredged and discharged at the other side of the harbour entrance to prevent it silting up and to reduce erosion on the north side. This system had been installed successfully at Vizakhapatnam, India, in 1935 after discussion with RPT. Studies were carried out at the Hydraulics Research Station at Poona and the experimenters were able to persuade the government that a layout as at Madras would, in their view be superior. RPT revised the layout to include a sandpump offshore of the main breakwater to enable the accumulating sand to be pumped across the harbour. The harbour was constructed, but it was many years later before the authorities installed the sandpumps to enable entrance depths to be maintained and erosion on the downdrift held in check.

The firm has been responsible for port works in Libya at Derna, Benghazi, Marsa Brega, Misurata and Ras Lanuf. The last of these at the bottom of the Gulf of Sirte serves an oil refinery and petrochemical works which together with a township has been constructed at the edge of the vast barren desert. Visitors, on the four hour drive from Benghazi on the new coastal highway, still pass the notices at 'Marble Arch' and elsewhere warning of mines left by the armies of the second world war which fought in the arduous conditions back and forth along this coastline.

Photograph of a model of Benghazi Harbour, Libya.

The port includes some 3 km of coastline and encloses 430 hectares. Two rock filled breakwaters, the largest of which is 2.5 km, are faced with tetrapods of up to 46 tons to break the force of waves in excess of 8 m high which occur in winter storms. The rock for the breakwaters and all the concrete had to be quarried selectively from a thin layer of cap rock on the

Ras Lanuf Harbour, Libya, is a new deepwater harbour to serve the export and import requirements of a new industrial complex. The photograph shows the launching in March 1982 of one of the many concrete caissons used to construct one of the two breakwaters.

nearby desert floor. That operation extending over almost two years saw heavy dump trucks passing into the port area day and night, the intervals between them being less than one minute over long periods.

The cargo berth is formed of 21 concrete caissons each weighing 5500 tons. They were cast on the pumped sand reclamation and then 'tipped' into the sea to be floated into position by the careful dredging away of the sand; an unorthodox and spectacular method but entirely successful and economic.

RPT has a reputation for strict control of site operations.

Inspite of the severe logistic problem occasioned by the remote site and tight timescale the RPT site supervision team was able to secure from the Korean contractor, Hyundai and its Philippino labour force, some of the finest quality construction work seen anywhere. RPT has a reputation for strict control of site operations which is amply justified by the work at Ras Lanuf.

David Hookway who was in charge of the Ras Lanuf project had been in charge of RPT's operations in Brazil in 1971 to 1977 where he had carried out major works of a very different nature.

RPT started work there under a project financed by the UK Overseas Development Administration to prepare Master Plans for the ports of Recife and Rio de Janeiro. The latter port, which spread as a narrow strip along many kilometres of the coastline of the magnificent sheltered bay in which it lies, was by then hemmed in by major road projects including elevated roadways constructed to deal with the massive traffic of that city. Hookway, who rapidly became fluent in Portugese, is credited with having brought together, almost for the first time, the officials of the ports, roads, railway and other authorities to try to evolve a coherent development programme.

In 1975 RPT and its local associate Planave were responsible for the World Bank financed National Ports Study. This involved a nationwide survey covering 29 ports, reviewing traffic forecasts and facilities leading to the development of master plans, investment strategy and conceptual designs particularly for container handling. The complete report occupied many feet of shelf.

Thereafter RPT prepared drawings and documents for the Sepetiba Marine Terminal. This included a 22 km access channel, 3.5 km conveyor, two coal berths, two iron ore berths and 1.5 million m^3 of storage. Mike Kormanicki, a relatively junior member of the RPT team, was appointed by Planave to supervise the construction. When the RPT office was closed as foreign currency became scarce he remained in Brazil to super-

Container Terminal, Clydeport, Greenock — a new container quay, 260 m long with a potential dredged depth of 15 m. Included in the project were mooring dolphins, port buildings, floodlighting and a 9 hectare storage area.

vise other projects for that firm until he returned to the London Office in 1987.

In 1983 to 1986 RPT and the Economic Studies Group played a major role in assisting the Mexican Government in implementing development plans for the four ports at which it was intending to focus its industrial development. The object was also to reduce pressure on the vast sprawl of Mexico City itself and the other two major industrial cities on the plateau. The project, financed by the World Bank, included studies of land usage and transport, traffic forecasts and financial analyses. Again the staff had to develop fluency in Spanish in order to work effectively and live in the community.

At Greenock on the Clyde the firm provided a feasibility study, Parliamentary plans, site investigations, design and supervision for a new container terminal. It was built on the site of the old Princess Pier and Albert Harbour which were closed due to changes in the pattern of both passenger and cargo traffic. The berthing face is 1200 ft long, sufficient for two vessels and the quay is equipped with two 35 ton container cranes. This project was the first occasion upon which the BSC composite section Unissen piles were used. The new docks were opened when the M. V. *American Resolute* docked in a force 8 gale in June 1969 inaugurating a weekly service to New York and Baltimore.

Aberdeen Harbour, with which RP has had a long association. Followi the long term development plan, projects included the deepening and widening of the harbour entrance a the rehabilitation of six quays. A n ro-ro terminal for the Aberdeen-Lerwick Ferry was constructed wit the port.

In 1974 extensive new works were completed at Grangemouth on the River Forth. These included two new oil jetties, dredging and reclamation, the widening of a cut between two docks and a lock, lead-in jetties and impounding works. The first ship entered the new lock on the 10th Anniversary of the firm's first report entitled 'Proposals for the

Grangemouth Docks Development. The illustration shows the new lock of reinforced cellular monolith construction, complete with lead-in jetties and three pairs of mitre gates, which was built alongside the existing lock.

The opening of the new entrance lock at Grangemouth by HRH the Duke of Gloucester, 26th September 1974, when the tanker 'British Forth' passed through the lock.

development of Grangemouth Docks'. Although by no means typical, this very long gestation period is indicative of the knowledge span and project support required today to operate successfully a large engineering consultancy.

Finally, the firm undertook extensive works for the Aberdeen Harbour Board. The staged development of the original port included deepening and widening the entrance and the delicate task of providing open quays in front of the existing old masonry quay walls to stabilize them as the impounded docks were converted to full tidal operation as required by the offshore oil industry which operates round the clock. This work enabled Aberdeen to be the major port servicing the North Sea oil developments. A new ro-ro terminal operable at all states of the tide was also constructed within the existing port and a new fish market was provided. The firm's services included the preparation of the long term development plan, design and contract drawings and supervision of construction.

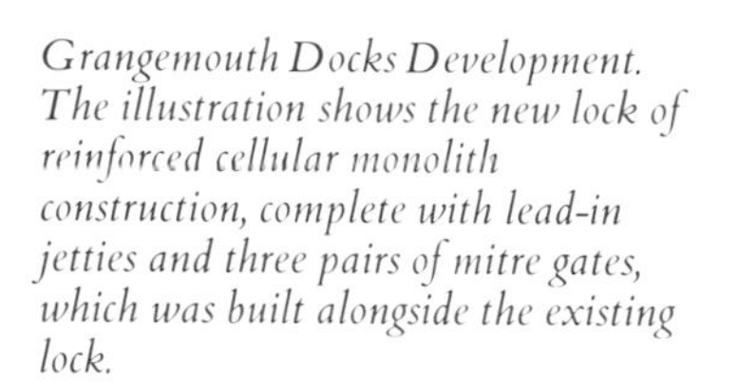

Ironbridge 'B' power station is a major 1000 MW coal-fired power station, sited on the banks of the River Severn, about 2 km upsteam from the 200-year-old Iron Bridge which symbolises the start of Britain's industrial age.

For over 50 years RPT has been involved in civil engineering work associated with power stations. Initially they were concerned with feasibility and geotechnical studies linked with the design of heavy duty foundations and cooling water systems. After the war, in association with the Swedish firm VBB, they extended their interests in power station work to India and Pakistan. During the past 25 years the firm has studied and reported on over 40 potential power station sites in the United Kingdom for the Central Electricity Generating Board (C. E. G. B.). Recent work has also included major schemes in Libya, Sudan and the Middle East.

During the past 25 years the firm has studied and reported on over 40 potential power station sites in the UK.

The Berkeley nuclear power station in Gloucestershire was completed in 1962. The firm was retained to produce reports dealing with siting, layout and foundation conditions, cooling water schemes and design specifications for all the civil engineering work.

In Shropshire, work commenced in 1963 on the construction of the 1000 MW coal-fired Ironbridge 'B' power station about 1½ miles upstream from the 200-year-old iron bridge which in many ways symbolises the birth of Britain's industrial age. The firm was responsible for the civil and structural engineering works including the superstructure, a 204 m high chimney, road and rail access, two bridges, the foundations and the cooling water systems. A Certificate of Commendation was awarded for the careful consideration given to the architectural aspects of the project.

The design of water cooling systems has perhaps become the firm's speciality, whether they be for conventional fossil fuel or nuclear power stations. Inland stations use cooling towers, but coastal stations create special problems involving means of preventing the recirculation of the cooling water. RPT engineers devised a number of interesting and novel solutions to the problems.

South Denes coal fired power station at Great Yarmouth in Norfolk was built on the foreshore between the sea and the River Yare. In its last stretch before debouching into the sea, the river turns abruptly and follows the coastline to the south until it reaches Gorleston. Research has indicated that this change of direction is comparatively recent (post-Roman) and was caused by littoral drift building up a confining bank of shingle and sand on the seaward side. All danger of recirculation was avoided by siting the station on the foreshore bank, with the cooling water taken from the river and the discharge to the sea. Water pressure in the

South Denes coal-fired power station at Great Yarmouth, Norfolk.

granular ground formation, however, was considerable and liable to tidal fluctuations. The level was held down by a de-watering well-point system until countered by the imposed weight of the station buildings. Another unique John Cuerel feature was that the foundation consisted of a concrete raft continuous over the whole area of the building without any movement joints or plinths for the column bases. Where the 30 ft high wall of the coal store had to cross recently excavated trenches innovative use was made of chemical consolidation to provide the requisite bearing strength.

At Aberthaw on the Severn, the solution to the cooling water problem was to provide an intake tower in the form of a caisson set in the river. With the intake at low level and the outlet on the surface, vertical separation was provided. The caisson was floated into position and sunk on to a prepared foundation. Thereafter a careful excavation connected the tunnels for the cooling water upwards into the caisson. At Hinkley Point and Oldbury nuclear power stations, both on the Severn, water depth was insufficient for this system. A barrier was constructed in the river opposite the power station frontage at Hinkley Point, permitting flow round the ends into the cooling system, and the outfall was discharged through a tunnel beyond the barrier. The system at Oldbury was based on a similar principle, but a lagoon was formed by walls on the rocky foreshore, with outfall tunnels to a deeper river channel beyond. The original concept was that there should be gates in the wall for filling, but the C. E. G. B. objected, fearing that they may become jammed with floating logs, and the wall had to be made solid. The determination of the optimum height of the wall to maximise containment and yet be low enough to overtop at all high tides called for considerable research.

Aerial view of Oldbury 440 MW nuclear power station showing cooling water reservoir.

The Fawley 2000 MW oil fired power station in Hampshire is built on a 65 hectares site beside Southampton Water and extensive reclamation work was necessary before construction could start. Over 2 million cubic metres of sand and gravel from channel dredgings destined by the Port Authority to be dumped at sea were diverted to fill the site and to provide aggregates for the concrete. There was a requirement that Southampton Water should not be heated by the outfall of cooling water, which necessitated it being taken through a tunnel to discharge in the Solent. Soil conditions severely limited the choice of the route which was further complicated when the C. E. G. B. requested that the tunnel should be dri-

The Fawley 2000 MW oil-fired power station near Southampton.

ven on a rising grade so that it could be drained back into the station. Another requirement introduced about this time, was that in order to minimise environmental impact, the elevations of power stations should be kept as low as possible. At Fawley, this involved setting the basement in the bed of Ower Lake, which was in the flood plain of Southampton Water and where the foundation conditions were indifferent. The turbine manufacturers advised that no deflection was permissible in the shaft of the machines, necessitating the eight underfloor inlet and outlet culverts being made monolithic so as to act as rigid cellular foundations for the full length of each set. The ground under the chimney foundation varied considerably from one side to the other, raising fears of differential settlement caused by the wind loading on the massive shaft of the chimney. In order to prevent this, an unusual device was adopted. A ring of sheet steel piles was driven around the periphery of the chimney base. Within it, but not connected to it, was constructed a circular cellular base with radial compartments which were filled with saturated sand and gravel to pre-load the foundation. As the chimney shaft was built, material was removed in stages, with the facility for correcting for tilt, should it occur. On completion, the total load was still in excess of the weight of the structure on the premise that the soil would be pressed down inside the steel sheeting ring like a piston acting in a cylinder. Only then was the base connected to the sheet piling thereby providing an additional resistance couple and mobilising a considerable mass of soil as a safeguard against earthquakes. An earthquake actually occurred shortly after completion of the project, having its epicentre only 15 miles off the Isle of Wight, without any ill effects.

An earthquake occurred shortly after completion . . . without any ill effects.

A further problem arose at Fawley when the C. E. G. B. were refused permission to take the main transmission cable from the station across Southampton Water by overhead line, which meant that a tunnel had to be constructed under the waterway. This was problem enough with the unfavourable ground conditions, but further complications were to come. Firstly, the cables required to be cooled, and the heat discharge when fully developed was given as the equivalent of a 3 kW fire for every foot in length, and there were several thousand. The solution devised was to cascade water in the tunnel over a series of weirs in the trough in which the cables lay, the discharge being on the Fawley side. The critical decision

The 2000 MW base load power station at Pembroke in South Wales. The firm developed a new system for design and construction of the 214 m high concrete chimney which allowed rapid construction without extensive temporary works.

was on the depth required for the Fawley shaft in order for the tunnel to be able to pass safely under the deepest section of the bed in the middle of the waterway.

A further complication was that the Harbour Authority would not allow boring to be undertaken in this critical width of some 200 ft, because of its regular use at that time by the 'Queen' liners. The tunnel was driven from the Fawley side under maximum pressure of compressed air and the system is believed to be unique. When the mid section of the channel was reached the screws of vessels passing overhead could be heard quite plainly.

At Pembroke in South Wales, the firm was retained to carry out site surveys and geotechnical investigations, full civil and structural engineering designs and specifications and supervision of construction. The main station buildings were located in a tidal lagoon and initial works were carried out behind a rock dam. The cooling water discharge was through a 4 m diameter tunnel discharging 250,000 cubic metres per hour in a different direction from the intake. The firm developed a new system for design and construction of the main 214 m high concrete chimney which allowed rapid construction without extensive temporary works.

With the increasing adoption of nuclear power by the C. E. G. B., RPT became more and more involved with the civil engineering aspects. The

Eggborough power station, South Yorkshire — a 2000 MW base load coal-fired station designed for reliable long term operation.

earliest stations had steel pressure vessels which needed an external concrete biological shield surround to trap the neutron discharge. The heat exchangers were situated outside the pressure vessel, but were also radio-active and so required a secondary biological shield beyond them. With the increasing size of units, problems started to arise with the welding of thicker steel pressure vessels until the logical outcome came to be accepted that it would be economic to utilise concrete to form a prestressed pressure containment vessel which would also act as a biological shield. The additional attraction of this change was that if there should be an uncontrolled increase in gas pressure the steel vessel could fail explosively, whereas a concrete container would crack and allow some escape of gas, whereupon the internal pressure would be reduced and the prestressing forces would tend to close the cracks. The Oldbury nuclear power station on the Severn, the first to employ prestressed concrete reactor vessels, was built in the period 1962-64.

Oldbury Power Station was the first to employ prestressed concrete reactor vessels.

The development of these ideas for the replacement of steel with concrete was greatly assisted by the application and extension of Joe Otter's mathematical concept of Dynamic Relaxation to the stress analysis of prestressed concrete pressure vessels. Francis Irwin-Childs and Professor Alan Ross were appointed by the South of Scotland Electricity Board to advise them on the implications and acceptability of these developments with particular reference to the relationship between operating pressures and structural cracking. John Holland who had worked on the development of the design served seven years with the S. S. E. B. as the Appointed Engineer responsible for this work during the construction of Hunterston B nuclear station. He later re-joined the firm and continued working on nuclear safety problems and quality assurance.

Before leaving conventional power stations mention must be made of the 2000 MW Eggborough coal fired station built in the valley of the River Aire near Knottingley in South Yorkshire in 1969. The emphasis in the C.E.G.B.'s design criterion was the need for reliable long term operation. RPT was responsible for the siting studies, site investigation work, full civil and structural engineering design, assessment of tenders and supervision of construction. The station includes a 70 MW gas turbine installation for starting up and peak lopping operations, cooling towers, a 400 kV switch house and a concrete chimney 650 ft high. RPT was also responsible for the rail installations, coal handling plant and ash

The Gale Common Ash Disposal Scheme. The largest scheme of its type in the UK, Gale Common provides for lagooning 1 million tons of power station fuel ash each year.

disposal scheme. The power station draws its make-up water for the indirect cooling water system from the River Aire and the firm carried out careful environmental studies to protect the river regime.

The disposal of pulverised fuel ash is a major problem associated with coal-fired power stations. The Gale Common Scheme, the largest of its type in the United Kingdom, provides the solution to this problem for two larger power stations at Eggborough and Ferrybridge on the Yorkshire/Humberside border. Started in the mid-1960's, the project is due to continue until the year 2020. The scheme provides for lagooning 1 million tons of fuel ash each year. The ash, pumped as slurry, is stored in two main lagoons covering an area of 21.5 hectares, encircled by embankments composed of fly-ash and quarry shale. The total length of the embankments is 8 km and, when completed, the lagoons will form a landscaped artificial hill 49 m high, containing 15 million cubic metres of ash. An earlier smaller scheme was the Devil's Dingle Ash Disposal Scheme for the Ironbridge power station in Shropshire for which RPT was also responsible. The ash was deposited in a lagoon formed by damming a valley with an embankment constructed progressively in stages, partly of clay and partly of compacted PFA. The scheme was completed in 1985, and eventually the dam and lagoon surface will be soiled, grassed and landscaped.

The firm participated in a comprehensive study of the power and water requirements of the whole State.

Probably the most important overseas power station contract in recent years has been that undertaken for the Ministry of Works, Power & Water, Bahrain. The 100 MW Sitra gas-fired steam turbine power and water desalination station was built on reclaimed land 1 km offshore in Bahrain. The firm participated in a comprehensive study of the long term power and water requirements of the whole State, which incorporated detailed proposals for the phased construction of the station.

The 100 MW Sitra gas-fired steam turbine power and water desalination station on reclaimed land 1 km offshore in Bahrain.

Since the second world war the firm has been engaged in the design and construction of over 2000 miles of oil pipelines together with their pumping stations, storage depots and communication systems. In the Middle East a route for a 34 ins/36 ins diameter oil pipeline from the head of the Persian Gulf westwards to a terminal on the Mediterranean coast was undertaken for Middle East Pipelines Limited. Starting in the Euphrates valley the route crosses the Iraqi and Syrian deserts, passing close to the ancient city of Ur of the Chaldees and the Roman ruins at Palmyra in Syria. Rising to over 2000 ft in Syria, the route descends through the Homs Gap, between Lebanon and the Alawi mountains, to the Syrian port of Tartus. The route is some 700 miles long. Surveys were also made for a refinery site and an oil harbour at Tartus and a scheme prepared for fourteen protected oil-loading berths for 32,000 ton tankers with an entrance channel through the Ruad reef. The scheme was designed to handle the products arising from the refining of 20 million tons a year out of the total proposed through-put of 33 million tons a year. Sadly, due to the political instability of the region and the advent of super-tankers, the pipeline is not used today.

In 1962, the firm completed a £25 million scheme for the National Iranian Oil Company. The pipelines distributing oil products from the Abadan refinery to Tehran, with extensions to the Caspian, to Meshed and to Isfahan consisted of 1500 miles of pipeline, 16 pumping stations, (some diesel and some electric), tankage and allied facilities. From sea-level at Abadan the pipeline rises to 7300 ft between Abadan and Tehran and to 5100 ft between Tehran and the Caspian Sea which is 80 ft below sea-level.

In 1971, the East African Harbours Corporation commissioned the firm to design and supervise the construction of a pipeline and offshore single buoy mooring at Dar-es-Salaam. Inspite of the extreme difficulties

The pipeline and single buoy mooring designed for the East African Harbours Corporation, at Dar-es-Salaam, Tanzania.

of working in monsoon storms, the work was completed and officially commissioned by President Nyere in July 1973. The first vessel to use the buoy discharged successfully 55,000 tons of crude oil destined for the Ndola refinery in Zambia.

* * *

Rendel Palmer & Tritton's entry into marine work came in the 1880's with the design of river ferries as an adjunct to the firm's extensive railway work in India. Harbour craft soon followed together with a fleet of small craft for the Uganda Railways for use on Lake Victoria. By the 1920's a wide range of craft including tugs and barges, floating cranes, despatch and support vessels, launches and river steamers had been designed and their construction supervised by the firm, for clients all over the world.

Following the conclusion of hostilities after the second world war and with independence being gained by former colonial countries, the firm found itself heavily engaged in port craft replacement programmes, principally for Calcutta.

Containerisation had not arrived and Calcutta was seen then as the major port for import and export for the sub-continent of India.

Large drag suction dredgers to the firm's designs were built to keep the channels of the sand bars in the Hooghly estuary open, as well as bucket dredgers and attendant barges for the maintenance of the impounded docks system.

Shallow draft, centre-well, trailing suction hopper dredger Subarna-Rekha, designed for maintenance of the upper reaches of the River Hooghly, India.

Other ports were not forgotten in the 1950's and craft to the firm's designs were completed for Karachi, Chittagong, Madras and for Takoradi in newly independent Ghana. Nearer home, a dredger for Limerick, Eire was a notable 'first' for the firm, being a combined grab and trailing suction hopper dredger of which very few, even today, have been built.

Light vessel 'Candle' designed for Calcutta Port Commissioners.

The 1960's saw intense activity in replacement of the Calcutta fleet of port craft with floating cranes, tugs, research vessels and pilot vessels. Calcutta's move away from steam driven machinery had been somewhat delayed due to the war. The multiple grab hopper dredger built at Henry Robb's shipyard at Leith was among the first of the new diesel powered craft to enter the service of Calcutta Port Commissioners.

The hopper dredger *Tung Hai* and grab dredger *Nan Hai* were built in 1964 for the port of Keelung, Taiwan, the latter in the port workshops to

'Tung Hai' — a trailing suction hopper dredger for maintenance dredging in Taiwanese ports.

full working drawings provided by RPT.

About this time the firm was becoming involved in design of research vessels, principally through the World Bank's continued interest in maintaining Calcutta port. A launch and a much larger vessel were designed and built for the work of the newly constituted Hydraulics Research Department of the port. Through this association, the MAFF laboratory at Lowestoft, UK, invited RPT to assist in the design of a small fishery research vessel. The *Corella* proved to be a success in service and following on a few years later the firm designed a much larger ship for world wide operation. This vessel, the *Cirolana,* has proved to be one of the

'Corella' — a 40 metre stern-trawler type vessel designed to carry out hydrographic, marine biological and fisheries research in UK coastal waters.

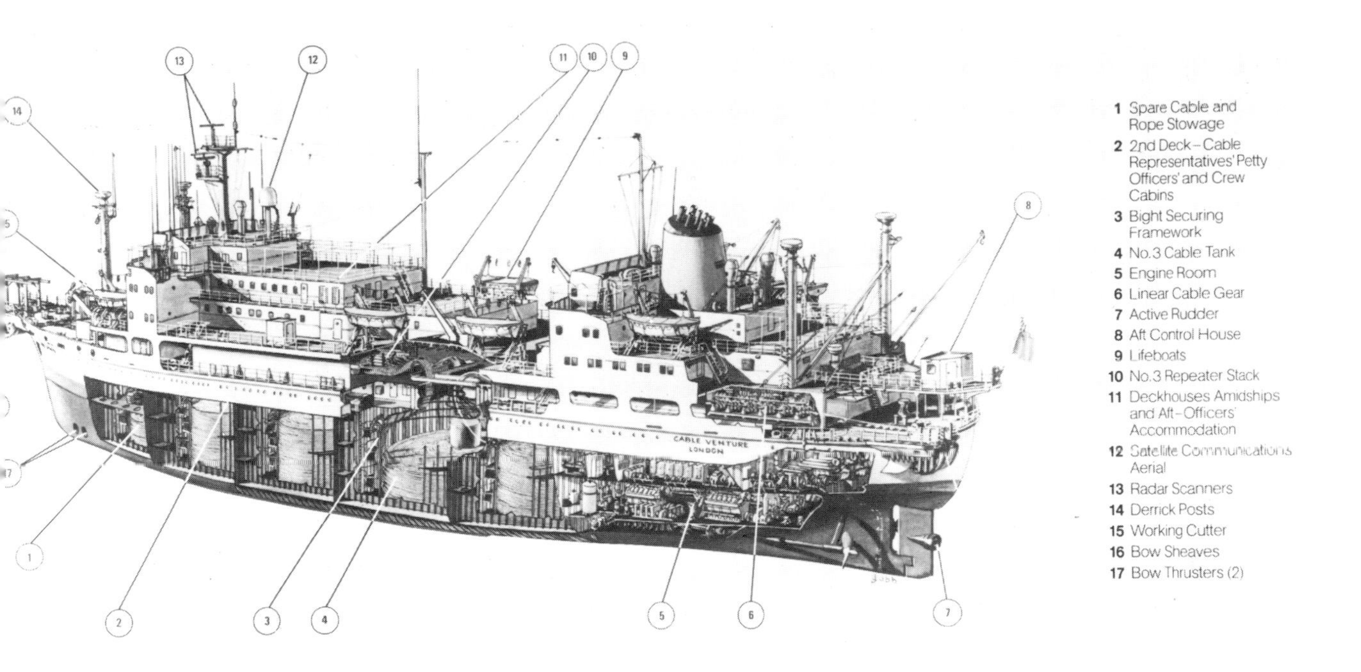

This vessel, built in Germany as the 'Neptun', was designed as a combined bulk carrier/cable layer. The vessel underwent a major conversion to take its place as flagship of the Cable & Wireless fleet as the 'Cable Venture'.

most successful of her type and has been copied by a number of other countries for their own research needs.

Another productive phase in the marine sphere at this time was work carried out for Cable and Wireless plc on the cable ships. The appointment of RPT initially covered stability investigations at the time of vessel refits but this led to designs for major modification to facilitate deployment of a range of new cable equipment. This involvement occurred in the 1970's when the major reference was a cable full clear working deck and deck-house on the ex-German cableship/bulk carrier *Neptun,* now renamed *Cable Venture,* which is still the largest cable ship in the world.

There was also a change in emphasis about this time when RPT was asked to supervise construction of vessels which had not been designed by the firm, the first of these being *Sir Thomas Hiley* a modern diesel electric suction hopper dredger for Brisbane, Australia.

The firm had always been able to function in a 'trouble shooter' role and this led to its involvement in dredging plant for winning material at sea for the construction industry.

Another interesting development arising through an association with the Charles Hill shipyard in Bristol was modification to two existing ships to enable the bulk transport of Guinness. The firm was associated with a pioneer scheme to ascertain if such sea transport was feasible, as opposed to the existing refrigerated cargo ship methods. Success with the experiment followed on with the design and construction of the *Miranda Guinness* in effect a fully equipped floating brewery.

In the 1970's the advent of containerisation was in full swing and the dredging fleets of the port authorities were being phased out due to the physical move of port facilities to deeper water. Although basic port craft design work was still pursued, the firm's Marine Department found itself moving more towards operational and feasibility studies. A major study carried out in 1978 was the Suez Canal Tanker Stopping Study. This involved desk, model and simulator studies as well as full scale stopping trials with a 70,000 dwt tanker in the canal. The first phase widening of the canal was nearing completion and the ship the *Daphne* was used as a 'model' in the last unwidened stretch in a series of stopping manoeuvres. The firm took full responsibility for these trials even to the extent of providing a pilot and tugmasters from the UK.

Research vessel 'Frederick Russell'. Converted from the modern stern trawler 'Gravinez' on behalf of the Natural Environment Research Council.

The 1980s have seen port craft designed for clients in Trinidad, Jordan,

Iraq and Burma. Little work has however been won for UK ports but good relationships have been developed with the sand and aggregate dredging industry through the firm's work in the 1960's. New buildings and full scale conversions of small bulk carriers to dredgers have been carried out, the most interesting to date being the *Cambourne* for Civil & Marine Ltd.

'Salammbo 2' — a cutter-suction dredger for capital and maintenance dredging in Tunisian ports.

An increasingly important area of work lies in the field of operational and feasibility studies, frequently involving several of the firm's departments. In the period from 1960, the Marine Department has participated in over 80 projects of this nature. These include several shipbuilding and ship repair facility studies, vessel survey and modification projects, shipping line studies, port master plans, dredging studies, and oil industry related projects.

The Marine Department also provides a valuable support service to the maritime civil engineering activities of the firm. Floating stability, and behaviour under tow and during the transition stages of installation on site have been investigated for a range of dock, lock and barrier gates and for floating caissons for both structural and offshore installations. Advice on navigation in restricted channels, including the behaviour of vessels when steaming in shallow water, has been provided as part of the overall design of ports, as has the establishment of tug requirements for the safe berthing of vessels. The department is staffed by qualified naval architects and marine engineers able to deal with any project involving buoyant structures.

Ralph Downham — Partner 1970-85, then Director to 1988.

From 1968 to 1988 the Department was directed by Ralph Downham who was admitted into the Partnership in 1970. He is a qualified Naval Architect and Fellow of the Institution of Civil Engineers and Institution of Marine Engineers. Prior to joining RPT in 1966 he was Chief Naval Architect and Technical Manager at Cammell Laird's shipyard at Birkenhead and responsible for the design of a wide variety of vessels built by the yard. Latterly, he has been involved in checking the floating stability of various wave energy devices and responsible for navigation studies associated with the proposed Severn and Mersey barrages.

William Dell, naval architect and Assistant Director, has managed many of the Department's projects.

The post war oil and gas industry, especially that in the North Sea, has created a wholly new activity for the firm, and over 50 major projects have been undertaken throughout the world in the past 20 years. Not only have they been concerned with feasibility and economic studies and with the design and supervision of construction of oil and gas installations, but RPT have been retained to undertake the preparation of highly technical and sophisticated reports dealing with such subjects as the environmental, design and safety aspects of offshore structures in the North Sea, the repair and maintenance of steel jacket platforms, the safety of offshore installations against ship impact, and technical appraisals for major insurance claims consequent on the failure of offshore structures. When concrete gravity based structures were first introduced into the North Sea, RPT, in conjunction with Lloyds Register of Shipping, jointly developed the SEAFACS suite of computer programs for dynamic and fatigue analysis of offshore structures in random seas.

The conversion of the 'Lady Patricia' for Arthur Guinness & Son, involved the installation of a complete outfit of stainless steel beer tanks to replace the cargo handling system for portable tanks originally installed in the ship.

In Abu Dhabi the firm designed platforms for British Petroleum for the Zakum, Umm Shaif and Uweinat fields. These included various jacket structures in water depths up to 37 m on a range of foundation strata, and numerous design and audit appraisal assignments. Extensive work was undertaken in Brazil for Petrobras, including the design of a family of jacket structures of modular construction for water depths up to 70 m. In Norway they carried out extensive work in British Petroleum's Ula field

BP's Ula platform complex in the Norwegian sector of the North Sea. RPT provided independent design verifications of the three production, drilling and quarters jackets, decks and selected module structures.

including independent design verification of jackets and deck structures for production, drilling and quarters platforms. At home they undertook analysis of moorings in the Brent field for Shell Expro and undertook the preliminary design for BP's Southern North Sea Development in the Cleeton and Ravenspurn fields. These included three wellhead platforms and one wellhead tower with 18 slots in 44 to 47 m of water.

The Offshore Structures Department is led by John Dawson, who had joined RPT in 1968 and was appointed a director in 1985. During 1979/80 he was seconded to Petrobras in Brazil where he was involved with the work just described.

In 1987 RPT became a member firm of the Offshore Certification Bureau.

In 1987 RPT replaced Sir William Halcrow & Partners as a member firm of the Offshore Certification Bureau. OCB is currently the Certifying Authority for 9 fixed platforms in the North Sea, two jackup drilling rigs and one semi-submersible diving platform.

Rendel Palmer & Tritton completed two major projects associated with the requirements of the oil exploration industry in 1977. Both were on the west coast of Scotland, and each exceeded £12 million in value. At Hunterson Sands on the Ayrshire coast, where an extensive foreshore adjoins a deep water channel, the firm was associated with the design and construction of a large dry dock for building the Anglo-Dutch Offshore Concrete Platform. Measuring 150 m x 150 m x 15 m below high-water level, the dock was constructed in nine months by dredging some 1.75 million cubic metres of material which was subsequently used to provide the adjacent working area. The reclaimed site is joined to the mainland by a bridge and causeway. RPT was chiefly responsible for monitoring

Artist's impression of the dry dock at Portavadie, Scotland, built for the construction of large concrete gravity oil platforms.

construction progress, financial control and certification of costs. At Portavadie, at the entrance to Loch Fyne, RPT provided managing services and design assistance for another large dry dock and associated infrastructure for the Department of Energy. This dock was also for the construction of large concrete gravity oil platforms for use in the North Sea, and the project included a new permanent village for the workforce and their families. In recent years the site has been converted to a fish farm capable of producing 3000 tons of salmon a year. The conversion involved the blasting of 20,000 tons of rock on the western promontory to form a basin for four 25 m diameter plastic polymer lined fish tanks, the whole project being completed in a little over 60 days.

The firm was appointed as principal adviser to the Government.

In 1977, the firm was appointed by the Department of Energy as principal adviser to the Government in a major study to determine the economic and technical feasibility of power generation from the energy of waves. This involved the technical and economic assessment of devices prepared by other interested firms and organisations and the examination of all aspects of wave climate, hydromechanics, structural design, moorings and mechanical power take-off. The electrical side of the assignment was handled by consultants Kennedy & Donkin. The following devices were included in the study: the Salter Duck Device, the N. E. L. Oscillating Water Column Device and the Lancaster Flexible Bag, the Bristol Cylinder, Vickers, HRS Rectifier and the W. P. L. Raft schemes. Benign sources of energy such as waves warrant develop-

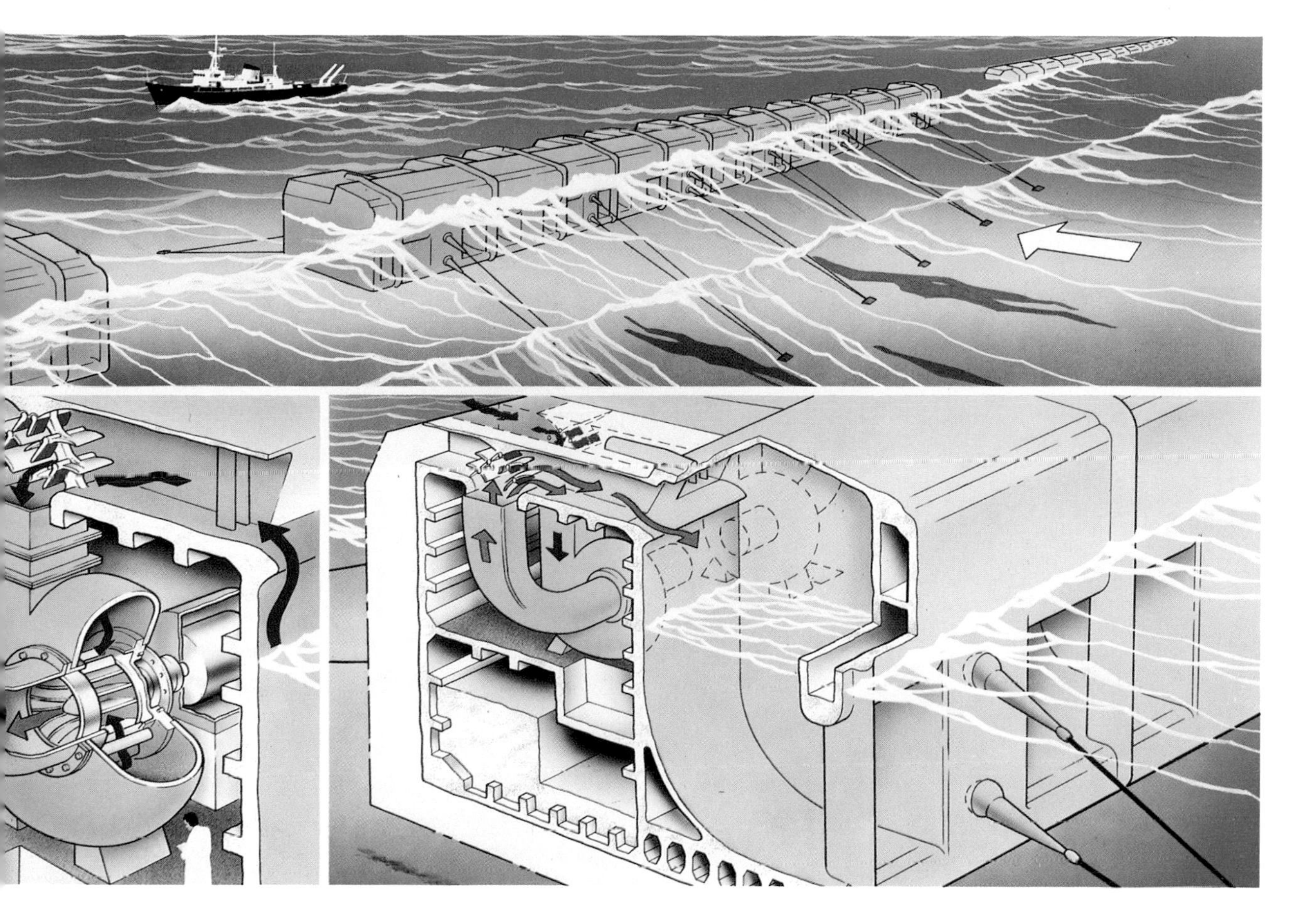

One of the many wave energy devices studied by RPT in its role as principal adviser to the Department of Energy.

ment study but the costs of maintenance and power transmission are high and, except in special circumstances they cannot yet compete economically with fossil fuel or even nuclear generation.

The firm has recently been asked to assess the structural design of two small-scale wave energy devices for the Department of Energy. The first proposal by Sea Energy Associates Ltd. is for a novel circular floating tubular steel structure supporting an air bag wave energy device. The second, smaller device was proposed by Queen's University, Belfast, and involves constructing an oscillating water column in a small gulley on Islay, in the Inner Hebrides. The Department of Energy have now approved funding for phase II of this project, involving construction of the civil works and assessment of their hydropneumatic performance.

* * *

This chapter has been able to touch briefly on the range of projects which RPT has carried out and continues to carry out throughout the world. However one major marine works project which has caught the imagination of the public generally and is a showpiece of RPT's activity has not yet been described. It is the Thames Barrier, constructed to relieve the capital city of the threat of a storm surge from the sea which so nearly came to devastate it in 1953. That project is described by Peter Cox in the next chapter. He took over in 1971 from Harold Scrutton, who followed Joe Otter, in charge of the project, and was responsible for the detailed design and construction phase. At the time of writing he is still negotiating the final fees with the client. There cannot be many professions, or even industries, for which projects run as long as they do for consulting engineers. Many clients nowadays 'shop' for consulting services ignoring, to their own detriment, the effort and skills which have to go into developing and maintaining the accumulation of knowledge which, integrated with current technology, is the life blood of a consulting practice.

... accumulation of knowledge ... is the life blood of a consulting practice.

10
The Thames Barrier

The severe floods in 1953 which devastated much low lying land along the East Coast and in the Thames Estuary caused RPT to be occupied on a project which would rank high amongst the most prestigious ever undertaken by the firm. The Thames Barrier was not in fact to be completed until some 30 years later but the studies leading to the final project, the design and the construction illustrate so well the range of factors with which a modern day consultancy practice has to deal that it is instructive to look at the whole saga.

. . . a project which would rank high amongst the most prestigious ever undertaken.

Archeological evidence indicates that at the time of the Roman Occupation of Great Britain the tide did not reach as far as London Bridge and that under normal flow conditions water level was about mean tide level.

The Bridge confined sea going vessels to the river downstream where tidal berths were developed. Banks were raised to protect the marshes and land behind from inundation by upland floods or the occasional tidal surge from the sea. Consequently the silt, which in time of floods had been deposited on the marshes thereby raising them, was no longer deposited. The land dried out and consolidated; and slowly the level of the protected land fell below the tidal levels in the river. The earlier walls and wharves were 15 ft lower than those built in recent years.

It is now known that in addition the general level of the south-east of England was slowly falling in relation to sea level. Thus the tides were progressively able to penetrate further up the river and historical records refer to floods from what we would now recognise as storm surges.

The inexorable rise of sea relative to land continued but it was not until after the severe floods of 1953 that a detailed study was made and a reasonably coherent scientific explanation to explain these events was able to be put forward. Levels recorded at London Bridge since the mid 18th century, when the first Ordnance Survey of Great Britain was made, show that extreme high tide levels have increased by about 4 ft in 160 years – say 0.3 inches or 7mm per year.

Sea levels have been rising relative to the land at about one foot per 100 years. This follows the retreat of the polar ice caps since the last Ice Age. Half of this is considered to be due to the general rise of the sea level. The other half results from the tilt of the tectonic plate on which Great Britian rests. Additional sinkage of the London basin results from the extraction of artesian water which has allowed the underlying strata to consolidate.

An artist's impression of what London would look like under flood.

The increase in depth of the Sea Reaches has allowed increased tidal penetration of the river. High water levels have been further raised by the

The completed Barrier during a test closure of all the gates.

compounding effect of the general dredging and smoothing of the profile of the upper tidal reaches which has taken place to facilitate navigation.

However another factor has to be brought into the equation to explain the exceptional high tides which occur in such a random fashion. It is a meteorological phenomenon, now regularly displayed on our television screens – a deep depression crossing the North Atlantic. This area of low pressure raises a hump of water, perhaps about one foot in height, a thousand miles in diameter and travelling at 40 to 50 miles per hour. If a depression passing north of Scotland turns southwards down the North Sea the funnelling shape raises the level further and the wind shear associated with northerly gales along the East Coast adds to the effect. Even at this stage a surge in the Thames Estuary is not certain as the weather pattern can carry the hump towards the Low Countries or the German coast line. In 1962 such a surge in the Hamburg area rising 14 ft above predicted levels claimed 312 lives and caused extensive damage.

Whether or not a surge will cause devastation depends on its phasing in relation to high tide. The 1953 surge tide did not occur at highwater but nevertheless rose at London Bridge 3.7 ft above the level of a high spring tide. Later the Thames Technical Panel, set up to study the problem, was to conclude that flood defences for London should protect against a surge tide some 6 ft above the 1953 flood level. That level adds the hump of the 1953 surge to a high spring tide, a probability which could occur once in one thousand years – but could occur in any winter as the random distribution of earlier surges indicates.

That the 1953 floods did not flood parts of London, as had the 1928 floods when 14 people were drowned, was because the overtopping of the defences along the East Coast and in Essex and Kent dispersed the

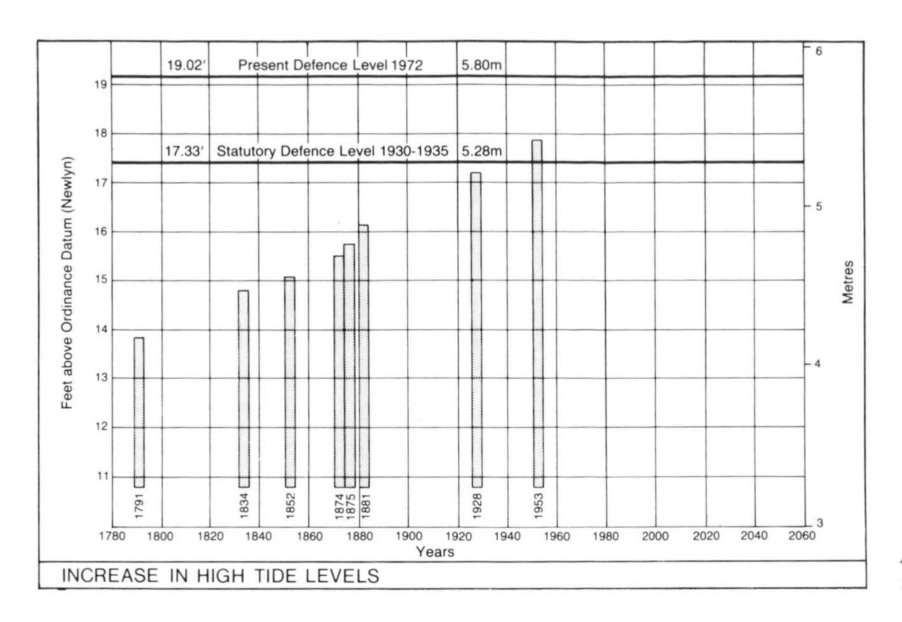

Diagram showing the gradual increase in high tide levels since 1791.

energy of the surge. There some 160,000 acres were flooded, over 300 people were drowned and great damage was done. Now with the estuarial defences raised the design surge would, in the absence of the Barrier, cause extensive and totally unacceptable flooding over wide areas of the capital with its highly developed technological services and valuable industrial and transportation infrastructure.

Raising the flood defences through central London, as was done unobtrusively as an interim measure, would have been an alternative, but protection against the long term design surge level would have cut off the river from view. Such defences would also have a low level of security and would be environmentally unacceptable, particularly now that the Thames is no longer polluted and its amenity role is widely recognised. (It was not until 1973 that the Water Act established the Regional Water Authorities and brought under the control of the new Thames Water Authority all aspects of pollution control.)

In dealing with the background of the Thames Barrier we have strayed from the main narrative but in so doing we have highlighted some of the fundamental forces of nature an understanding of which is essential for a successful engineering scheme. Other social and economic factors determining the final project will emerge as we proceed towards the commissioning of the Barrier in 1982.

It was not until the 19th century that action was taken to co-ordinate flood defences.

It was not until the 19th century that corporate action was taken to co-ordinate flood defences. After the two record tides of 1874 and 1875 Parliament empowered the Metropolitan Board of Works to have constructed flood defences to a specified level. Later the London County Council took over responsibility for co-ordinating the various public bodies with interests in flood prevention.

In the 1930s the LCC proposed a scheme for a barrage at Woolwich with locks for shipping. The approaching war with Germany caused this idea to be dropped as the structure was considered to be vulnerable to bombing. Barrages of this type had also been proposed in the 19th century and in the earlier years of the 20th. It was fortunate that those schemes never proceeded.

A barrage is a low dam across a river with sluices to control the flow of water, usually in one direction, and having locks for shipping. A barrier is normally open and is closed only for specific purposes, thus the flow of run-off water or tides and the movement of shipping is normally unimpeded. A barrage with locks would have impeded the movement of shipping, but more unfortunately it would have fundamentally altered the

These following pictures show a selection of the earlier schemes proposed for the Thames Barrier.

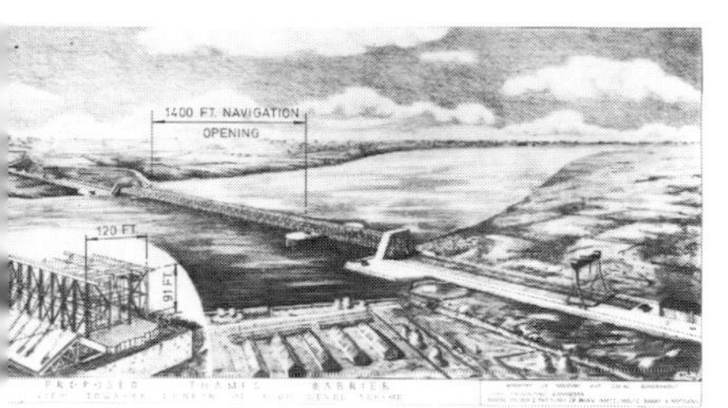

1965

regime of the river, almost certainly leading to extensive siltation as has happened elsewhere. This could have lead to the decline of London as a key centre of international cargo trade long before the change in shipping patterns caused this to happen.

The floods caused the Government to recognise the national importance of the surge tide floods. It set up a Committee under the chairmanship of Lord Waverley which reported in 1954. The recommendation affecting London read as follows:

> That the maximum standard of protection to be afforded by public authorities against flooding should in general be that sufficient to withstand the flood of January 1953, and this should be provided where flooding would affect a large area of valuable agricultural land, or would lead to serious damage to property of high value schemes such as valuable industrial premises or compact residential areas. Elsewhere the defences should be of a standard which would reasonably have been thought adequate before the flood of January 1953. In certain circumstances higher or lower standards may be appropriate. Anyone requiring a higher standard should pay for it himself.

1968

At this stage the Thames Technical Panel was appointed to devise a scheme which could be accepted by central Government and local interests. The Panel, under the direction of the Ministry of Housing and Local Government, comprised the Chief Engineers of the Ministry, other Ministries, the London County Council and other local authorities. Other organisations with an interest or able to contribute were also represented. An early recommendation was that consulting engineers should be appointed to carry out the detailed studies needed by the Panel before decisions could be taken. With the advice of the President of the Institution of Civil Engineers, Rendel Palmer & Tritton was appointed jointly with Sir Bruce White Wolfe Barry & Partners. This association continued through until 1972 when the Greater London Council, who had then become responsible for the project, decided to appoint only Rendel Palmer & Tritton to prepare the detailed design of the definitive scheme and to supervise the construction work.

The study period through to 1972 was long and complex. The joint consulting engineers worked closely with the client's directing group which was in turn commissioning further meteorological and tidal studies, and linking with the other interested organisations.

During this period Joe Otter, a Partner, devised the first mathematical model of tides in the Thames Estuary.

1969

Designing barriers and gates taxed the ingenuity of the consulting engineers and generated much discussion between them and with the governmental teams. Many schemes were prepared by them and proposed by others. (The latter included a canal through to the Bristol Channel to allow the surge to disperse there!) Sites considered in successive studies varied from Cannon Street in the City to Long Reach 18 miles downstream. Structures varied from a massive drop gate across a 750 ft navigational opening carried on 300 ft high towers, to a series of drum gates housed in concrete pits in the river bed which would have required excavation nearly 150 ft below high spring tide level. The high towers and massive suspended gate might even then have been considered environmentally unacceptable. The deep excavations in the river would have been hazardous operations.

The frontiers of technology were being stretched largely to meet the navigational requirements of the Port of London Authority. In 1954 the PLA were willing to accept in Long Reach a structure with two 500 ft openings and two 250 ft openings. Schemes to meet this requirement comprised a variety of drop, swing and retractable girders carrying sluice gates. By 1960 when the schemes were reported on the PLA had sanctioned other developments in the river and were demanding a 1400 ft clear opening.

The joint consulting engineers responded to this challenge with schemes which involve massive girders pushed from docks on either side of the river across the opening. Once in position gates on the girders were to be lowered to stop the flow of the water.

The 'low level' scheme produced by RPT had girders each 850 ft long by 77 ft wide and 70 ft high. Power demand was estimated at 5000 hp. The other consultants designed a 'high level' scheme having longer and larger girders and power demand of 7000 hp. The respective (1965) estimates of cost were £24 m and £41 m.

Each scheme had merits and disadvantages, which were expressed by the respective protagonists and the government hesitated to sanction such enormous expenditure on structures designed at the limit of technology.

As 1970 approached attention became focused on the Woolwich Reach as the preferred site. The PLA had recently announced the closure of the Surrey Docks and it was evident that the closure of the West India Docks would soon follow. Thus the numbers of large ships coming up the river were steadily decreasing. The Woolwich site had the great advantage that it was upstream of the lower entrance to the extensive Royal Docks and so would not interfere with ships then still going there. In the event the Royals closed to commercial traffic soon after the Barrier had been completed. London deep sea trade had by then moved downstream to Tilbury where the generous layout of the docks was well suited to facilities for the large containers of up to 40 tons capacity superseding break-bulk packages of up to 3 to 5 tons which had been the standard for general cargo for generations.

That land at Tilbury, which had been unused for many years, had been acquired by the PLA largely through the foresight of Sir Frederick Palmer when he was at the PLA. It is interesting to speculate whether, if that land had not been available, more pressure would have been added to the proposal to build a modern deep sea port as part of the extensive reclamation proposed at Maplin, at the mouth of the Estuary, to form the site of a Third London Airport (RPT also carried out for the British Airports Authority detailed studies on forming that vast site from sea dredged materials.)

Later RPT and Transportation Planning Associates were to become involved in the redevelopment of the docks under the aegis of the London Docklands Development Corporation. RPT was to be particularly responsible for the whole of the new infrastructure needed to enable the Surrey Docks area to be developed for new commercial and attractive housing schemes. TPA made traffic studies for the Docklands Light Railway.

Initially the Barrier at Woolwich was to have a main drop gate of 400/450 ft clear span. With the closure programme of the upstream docks then proceeding it was eventually possible to persuade the PLA and navigational interests to accept openings of 200 ft. This is the same as the opening of Tower Bridge through which ocean going vessels had passed safely during the 80 years that the Pool of London had been a thriving port area.

Even in the restricted length of the Woolwich Reach, alternative sites had to be considered and discussed with the local authorities and commercial interests. Environmental concerns and technical queries associated with the towers and the drop gates in rapidly moving tidal waters were causing the consulting engineers to favour the drum gate for this situation but they were still unhappy about the deep excavation needed.

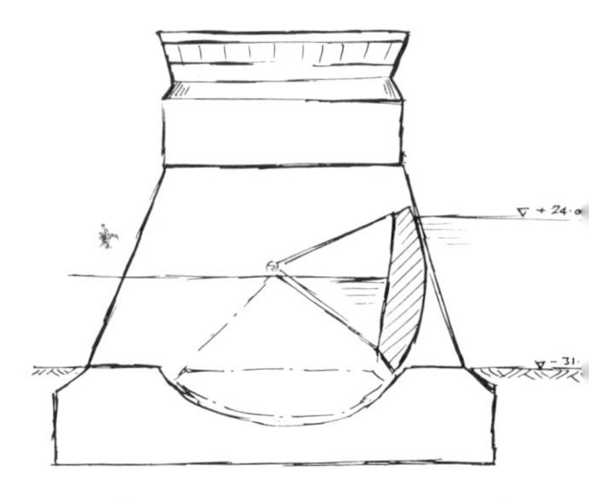

The first sketch, produced by an RPT technical engineer, illustrating the concept of the Rising Sector Gate.

RPT and TPA were to become involved in the redevelopment of the docks.

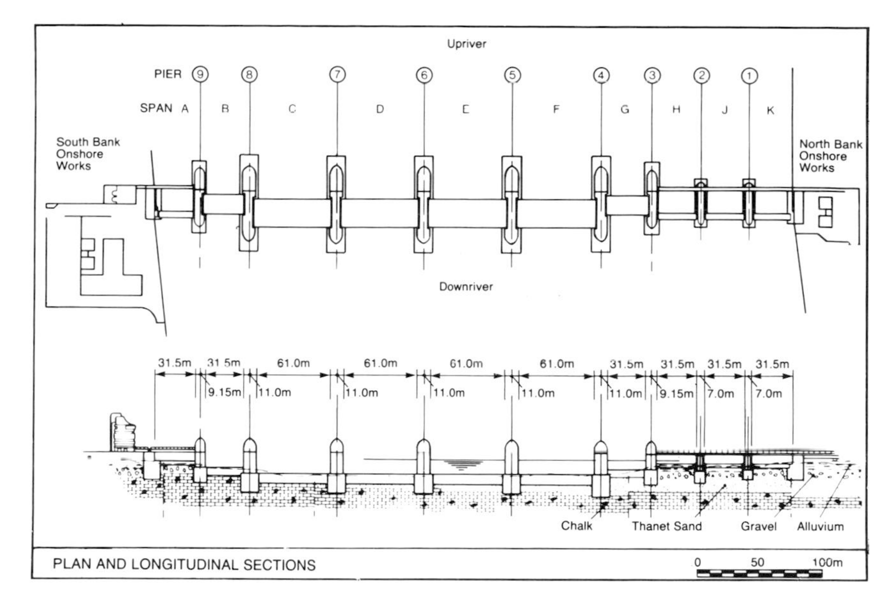

Diagram showing the plan and longitudinal section of the Thames Barrier piers and foundations.

The team working on the Barrier project had studied many forms of gate. One of the RPT technician engineers concerned, conceived the idea of the rising sector gate. This effectively combined a falling radial gate, used in many flood control structures, with the form of the drum gate, to provide a structure which required only a relatively shallow housing in the bed of the river. That simple sketch of the idea was to be developed by a top structural engineering team in RPT into a gate span 56 m long, 19 m high and 5.3 thick weighing 1450 tons, and supported at each end by 24.5 m dia, 1.5 m thick rotating disks each weighing 1100 tons. The recess in the river bed became a prestressed concrete 'eggbox' 60 m, long by 27 m wide, by 8.5 m deep weighing 9000 tons before it was floated into place, sunk or to its supports on the piers and filled with sand. The Barrier comprised four such gates, with flanking similar gates with 31 m openings, and four more falling radial gates close to the banks to provide the necessary total waterway area.

In parallel the mechanical and electrical engineers were developing the operating machinery: the larger gates were to be operated by four 1.1 m dia hydraulic cylinders each capable of a thrust of 1450 tons.

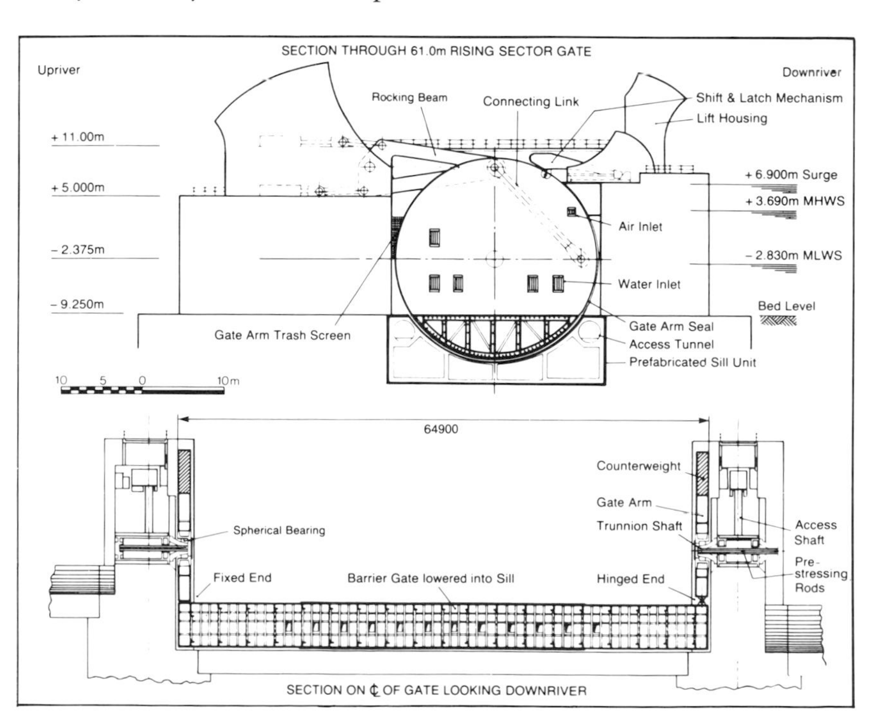

Typical arrangement of the Rising Sector Gates on the Thames Barrier.

In September 1971 the Greater London Council signed with RPT an agreement for RPT to design and supervise the construction of the Barrier at Woolwich. Thus, some 18 years after the floods which had started the search for a solution to their prevention the tempo of the work increased as RPT started on the detailed design of the final scheme.

In addition to the studies with which they had already been involved, RPT recommended that further detailed studies were needed to ensure that risks during construction and operation should be reduced to an absolute minimum. Many modern sophisticated techniques were used.

The manufacture of a trunnion shaft of one of the 61 m gates.

The welded box girder gate leaf was checked against the Merrison Rules which were being brought into use following disastrous failures of two notable box girder bridges (RPT had participated in the development of these rules). A 1:6 scale model of the leaf (10 m long) was tested to destruction at Imperial College, failing very close to the theoretically calculated load. Welding distortion studies were made to guide the contractors during fabrication of the gates. Hydraulic model studies were used to check that there would be no damaging vibrations as water flowed past the gate and to determine how to protect the river bed against erosion as the gate checked flowing water, or should a gate fail to close. The complex shapes of the shell roof structures required that the stresses in them should be determined from tests on a model in a wind tunnel. The massive trunnion shafts (stub axles) carrying the rising sector gate arms, each weighing 45 tons and carrying a load of some 5000 tons, were modelled using photoelectric techniques to determine the stress in the shafts and the bolts securing them to the piers. Over 80 boreholes were put down on the site to determine the ground conditions and the movement of water in the soils. Seismic studies were made to minimise the risk from earthquakes – small tremors occur in the UK more often than many people realise.

Clients are often reluctant to spend money on studies and research but there is no doubt that on the Thames Barrier project the willingness of the GLC to agree to these wide ranging researches paid handsome dividends during design, construction and subsequent operation.

The successful completion of the Barrier owes much to the working relationship established between the GLC and RPT. In the GLC a small project team co-ordinated the technical and administrative input from the GLC and the Ministry of Agriculture Fisheries & Food (MAFF) which was paying 75% of the cost. Political control on the GLC changed four times during the course, and that in central government also changed. Much of the project was carried out during a period when union and labour problems were the subject of intense national debate. Nevertheless, the consultants were always able to bring to the attention of the project team matters requiring decisions by the client and know that there would not be undue delay in a decision being given.

Drawings and documents for 28 main contracts were prepared by RPT.

Drawings and documents for 28 main contracts were prepared by RPT as part of an integral plan to suit the phases of the work. Co-ordinating these contracts was a major responsibility initially in head office and then in the site office from which the construction was supervised. Tenders for the three largest contracts – C3 Main River Works, C4D Machinery, and C6 Gates – were invited in 1973. Tenders for Contract C3 were received in November 1983 at the time when the negotiations between the Heath Government and the mine workers union were putting a strain on the whole economic scene. Not surprisingly it was not easy to finalise a contract which would be acceptable to all parties. Peter Cox, who had taken over the direction of the project from Harold Scrutton when he retired in 1972, was much occupied leading the RPT team in negotiations with the tenderers on behalf of GLC. The contract was finally signed in

ne of the two 3m diameter access tunnels hich accommodate duplicated power and ntrol cables and service pipework.

July 1974 with completion scheduled for late 1978. Labour problems added to the tough technical requirements of the project and there were major renegotiations of the contract in 1976 and 1978. At one stage MAFF was sufficiently concerned in reacting to the questioning on costs by the Public Accounts Committee to appoint an independent group of three wise men to review project control and finances. The Group was in due course named the Thames Barrier Advisory Team, the acronym TBAT having a less derisory sound! TBAT circulated amongst all the parties and, by and large, expressed themselves well satisfied with the management of the project at all levels.

Contract C3, for the main River Works, was the most evident of the numerous activities of the project. At the peak period some 2100 operatives and 400 staff were employed on the site, working three shifts around the clock. RPT had a matching site team of up to 70, controlling technical standards, programme and cost. The nine piers in the river were constructed in heavy steel sheet piled cofferdams, the largest being 77 m long, 17 m wide excavated to a depth of 28 m below high water level. At about that time Victorian sewers in some cities were collapsing forming

One of the sixteen 1.1m diameter hydraulic cylinders used to raise the main rising sector gates of the Thames Barrier.

large holes into which buses had fallen. A colloquial measure of the size of a hole came to be a double-decker bus (DDB) and on that scale the largest cofferdam was 440 DDBs. Large teams of divers were needed in the construction of a cofferdam, until it was possible to place underwater a 5 m thick concrete plug needed to stabilise the bottom and allow the cofferdam to be pumped out and construction of the pier to proceed. RPT also had engineers trained in diving for supervision. The volume of the largest plug was 6600 cu metres and its placing required continuous pouring for up to five days and nights. The construction work was a major logistic exercise as access to the piers was by either piled roadways from the shore or barges. Navigation had to be maintained at all times.

Perhaps the most delicate operation was the floating into position of the concrete box sills which formed the river bed housing for the 61 m gates. Clearances between the floating sills and the piers was only 50 mm. Hydraulic model studies were used to check the sequence of operations. Each three day operation was meticulously planned, rehearsed and directed from a specially constructed control room. This care paid off and the spectre of a sill jammed between two piers on a falling tide was happily avoided.

A precast concrete sill under tow.

Construction of the Gates started in Teeside in 1974 although, because of the slow progress of the River Works, it was not until 1980 that site erection was able to start. Each unit was brought to the Thames by barge and lifted into place by two 800 ton capacity floating cranes working in tandem. This delicate exercise required complete closure of the river for 12 hours on each occasion and a watchful eye on the weather. On one memorable occasion an optimistic forecast was relied on when the initial lifting preparations had to be made in the early hours of a December morning when strong winds and a blizzard required a courageous decision

The Thames Barrier during construction. At the top of the picture, two precast concrete sills are completed in the cofferdammed 'dry dock' prior to float out, and in the lower half, the installation of a gate arm takes place.

to proceed. All went well as, with the dawn, the wind dropped as predicted and the sun appeared. By midday the first gate leaf was in position.

Space precludes a description of the many other contracts which provided the extensive power, control and communication system; fire protection and security against sabotage; maintenance workshops; control buildings; and others as prosaic as the operational and maintenance manuals. The scale of the electrical services, for example can be judged from the fact that the central and local rooms required some 5000 relays to operate the 40,000 power machinery functions needed for the safe and secure operation of the Barrier. Security of operation was paramount in all aspects of the project. The project drew upon the accumulated experience of all the departments of the firm and required closely integrated team work.

All these contracts were carefully co-ordinated by RPT into the overall

Floating cranes moving the gate arm of a 61 m gate into position.

programme. The reward came in the early hours of 31st October 1982 when the first complete closure of the Barrier was achieved and it was declared operational. Those who had worked long hours to achieve the final link-ups, and others privileged to be present, recall with pride celebrating in a site office on one of the piers with champagne, cold eggs and sausages thoughtfully provided by an RPT team member!

Much work still needed to be done before the formal opening of the Thames Barrier by H. M. Queen Elizabeth on 8th May 1984, an event shared with many of those who had taken part in the project and members of their families.

In 1985, Rendel Palmer & Tritton, was honoured by the award of the Queen's Award for Technological Achievement for the firm's work in the design and supervision of construction of the Barrier. The Award was

The main control room of the Thames Barrier.

Aerial view of the completed Barrier during a test closure of all the gates.

presented by Baroness Phillips, Lord Lieutenant of London, at a ceremony in the Institution of Civil Engineers attended by some 500 members of staff and their families. All shared in the recognition of the team effort which had touched every aspect of the firm's activity.

Key senior engineers involved in the project included: Alex Young, George Carr, Lewis Coombes, successive project managers; Alan Mitchell, civil works design; Stuart Pratt, electrical services and controls; Geoff Miller Richards, Richard Tappin, gates; Bob Kirton, machinery; Paul Harris, hydraulics.

George Davies who was in charge of the RPT site team received the MBE.

Much has been written about the high cost of the project. When the design was finalised in 1973 the estimated cost was £110 m. The final cost

Inauguration of the Thames Barrier by HM Queen Elizabeth on 8th May 1984.

THE QUEEN'S AWARD FOR TECHNOLOGICAL ACHIEVEMENT

was around £440 m. The greatest part of the increase was due to inflation, which during the 1970's was well over 20% per annum in some years. The next major cause was low productivity and delay resulting from overt and covert industrial action, and the legacy of labour relations in the docks, which even now still colours some of the UK industrial scene. Construction difficulties account for some 10% of the increased cost and 5% is attributed to design development.

Now the Barrier and its tourist centre are a major attraction in the London scene, and it is good that a major engineering achievement is able to be so well displayed to the public. Most of the complex engineering, other than the massive arms which operate the gates, is hidden from view. The striking feature is the unusual shapes and glittering surfaces of the roofs which were conceived by the Architects' Department of the GLC which worked closely with RPT on all the architectural aspects.

The Barrier will still be operational well beyond 60 years theoretical design life. But by then future generations of RPT engineers may be devising equally ingenious schemes to deal with the increasing rise in sea level, above that allowed for in the design, which will occur if the more dire predictions of the 'greenhouse effect' come to be.

11
People and Organisation

In describing post-war projects and some of the general activities of the business, reference has been made from time to time to the particular role played by a Partner or senior staff. Because of the multi-discipline work of the practice those descriptions do not make it easy to discern the overall direction of the firm which was determining its future. This is therefore a convenient place in the record to draw together the complex network.

Ernest Buckton had retired as Senior Partner in 1947, very shortly after the completion of the firm's post-war reorganisation. He was succeeded by Ralph Strick, who occupied the Chair until 1961. In turn, Strick was succeeded by John Cuerel, the brilliant designer of Waterloo Bridge. Harold Scrutton, then came on the scene. He had joined the partnership in 1958, and took over as Senior Partner for the period 1966 to 1972. He was succeeded by Francis Irwin-Childs, a widely experienced and dedicated engineer who had been admitted to the partnership in 1961. He held office until 1978, when Peter Cox, the last holder of the appointment of Senior Partner before incorporation, took over.

All six men were very different characters. Buckton was, as we have seen, an autocrat and disciple of the Indian Raj. It has been said that he fostered a matriarchal influence in the firm, encouraging a monthly luncheon meeting for the partners and their wives at the St. Ermins Hotel in London. These were expected to be treated as a Royal Command and put the fear of God into several junior partners and especially their wives.

Strick had been 'discovered' by Buckton while working on the Bengal and Nagpur railway in India, and he tried to carry on the tradition, but the 'Hen Parties', as they became known, ended when John Cuerel took over. Buckton and Strick were the last of the senior partners having the old partnership philosophy laid down by Sir Alexander Rendel and carried on by Sir Frederick Palmer. Strick was small in stature and his tall Calvinist wife of Dutch-Boer extraction was long remembered by many members of the staff for always referring to her husband in public as 'my little tuppence'.

John Cuerel was undoubtedly a very brilliant engineer with an instinctive flair for design solutions. He rarely went overseas, and during his tenure of office, seventy per cent of the firm's work came from U.K. sources, whereas under his successor's leadership the position became reversed. He was basically a very shy man but he would express firmly his strongly held views. Nevertheless staff working closely with him were able to exchange their opinions on a mutual basis. When he visited a site a prime duty of the resident engineer was to locate for luncheon a pub serving his favourite Bass beer. London's elegant Waterloo Bridge will long remain a fitting memorial to his brilliance. Silvertown Way in Docklands, an early urban flyover, was built in the 1930's to his innovative

Harold Scrutton — Senior Partner 1966-72.

design and still remains in full use.

Harold Scrutton was very much a docks and harbours man. He started his working life with the Consulting Engineers, Walter Bridges. He then joined the Falmouth Dock & Engineering Company, engaged upon the construction of graving docks, jetties, workshops and oil installations. In 1929, he went out to Singapore on the staff of Sir John Jackson & Co. Ltd., who were engaged on the construction of the new naval base. He was employed by RPT for short periods in 1933 and 1938, and during the War saw service in the Engineers Branch of the Royal Marines. Upon demobilisation, he joined Edmund Nuttall & Co. Ltd., engaged on various tunnelling and drainage schemes, before returning to RPT for a third time in 1952. He became Resident Engineer on the Aden Oil Harbour project and was taken into the Partnership in 1958. He was instrumental in setting up Irendco, the firm's company in Iran, and through it developed the practice's important activity in oil pipelines and pumping.

Francis Irwin-Childs — Senior Partner 1972-78.

Increasing technical speculation meant that the firm needed to have in its team senior people who could be recognised as technical specialists, but who would not necessarily be considered suitable to take on the responsibilities of Partners. Scrutton established a grade of Associate and the names of the persons appointed appeared on the firm's note paper under that heading. (Associate Partner, a designation used by some other firms, was avoided because of the risk of the legal liability of a Partner being attached to such a title.)

In 1969 Scrutton started *Rendels News,* initially as a bi-monthly magazine. It was an excellent publication, eagerly awaited by both members of the staff and interested outsiders. For many years Eileen Jarvis edited the magazine which provided an intelligent commentary on the firm's activities, its past history and social news of interest to past and present employees.

Scrutton's successor, Francis Irwin-Childs, is a Norfolk man, and in his early days as a civil engineer saw service with the Norfolk, Buckinghamshire and Sussex County Councils. During the War he was employed by Sir Alexander Gibb & Partners as Resident Engineer engaged on the construction of underground factories. In 1945, he went out to the Gold Coast (Ghana) for the Ministry of Works engaged on the supervision of design and construction of railway bridges for the line from Dunkwa to Awaso. On the voyage to Takoradi he survived an air attack in the Bay of Biscay in which all the escort ships were set on fire and hundreds of lives were lost. Upon his return home he became Regional Production Engineer with the Directorate of Opencast Coal Production. He was then appointed to the Colonial Development Corporation as a Senior Engineer responsible for a wide range of development schemes in the Carribbean region. He joined RPT in 1952, and became a Partner in 1961. He was responsible for power stations, docks and harbours, and similar projects at home and abroad. He remained first and foremost an engineer and, today in semi-retirement, is still actively interested in major, sophisticated schemes such as the Severn and Mersey Barrages.

Peter A. Cox — Senior Partner 1978-85.

Peter Cox saw service in the Royal Engineers between 1942 and 1947, and then joined Lewis & Duvivier engaged on sea defences and marine work. He joined RPT's design staff in 1952, but left for a two year period in order to gain wider experience with Peter Lind & Co. Ltd., and Sir Bruce White Wolfe Barry & Partners. He became Resident Engineer for the construction of oil jetties in Iran upon rejoining RPT. Later, on his return home, he took responsibility for jetties, dry docks, and major harbour works both at home and abroad. He became a Partner in 1966 and Senior Partner in 1978. Upon the incorporation of the Partnership in 1985, he became Chairman of the new company until, on his retirement in 1988, he was appointed as Consultant. In 1980, he had the privilege of being elected President of the Institution of Civil Engineers. Not only has he presided over the successful completion of several major engineering achievements during his period of office, the most notable being the Thames Flood Barrier, but he also oversaw the traumatic events leading to the link-up with High-Point plc.

He was keen on engineers becoming more involved in wider affairs and to this end started the Infrastructure Policy Group of the Institution. He also served on the council of the Association of Consulting Engineers for seven years, becoming Honorary Treasurer. His interest in contractual matters took him on to the Committees which produced the 4th Edition of the ICE Conditions of Contract and the 3rd Edition of the FIDIC International Conditions of Contract. In retirement, he is Chairman of the Institution's Legal Affairs Committee.

Peter Cox was President of the Institution of Civil Engineers in 1980/81.

By 1960, all the Partners in the post Sir Frederick Palmer period had retired. It was customary for retiring Partners to be retained as Consultants.

Ernest Bateson retired in 1958 and remained a Consultant until 1961. Julian Tritton retired in 1955 and remained a Consultant until 1964. He was twice President of the Institution of Locomotive Engineers and is remembered for his erudite papers on the Limitations of Locomotives and the Standardisation of Locomotives and Rolling Stock. In 1951 and 1953 he was Chairman of the Association of Consulting Engineers. From 1955 to 1963 he was President of the influential International Federation of Consulting Engineers (FIDIC) whose Secretariat is based at Lausanne in Switzerland. His involvement with FIDIC arose from the post-war co-operation between RPT and VBB of Sweden, referred to elsewhere herein. That link and Tritton's influence was of great importance in the development of FIDIC. He was the Federation's signatory on the 1st Edition of the FIDIC International Conditions of Contract published in 1957. For his services to the Federation he was appointed an Honorary Member, an honour awarded to less than a handful of people. He was a charming and extremely popular man and is also remembered as one of the first people to instal a diesel engine in his private car. In 1929, he qualified as a pilot and was keen on flying as a hobby.

Julian Tritton was Chairman of the Association of Consulting Engineers in 1951 and 1953.

Alfred Clark retired in 1958 and in the same year Frederick Greaves was tragically killed in a yacht accident in Sweden.

The intrepid John Palmer, whose exploits are described in other chapters, retired in 1966, but remained a Consultant until 1980. In addition to his extensive overseas work he was mainly responsible for RPT securing the work in modernising many of the major UK ports to suit the requirements of post-war shipping. It was said that one stage the shelves of the National Ports Council were dominated by the green bindings of the reports and documents prepared by RPT! (Green was then the firm's house colour.)

Joseph Otter joined RPT in 1926, became a Partner in 1956, and died in harness in 1969. From 1937 to 1955 he worked in India first on temporary leave to Merz & McLellan and then as Chief Engineer and Manager of Merz, Rendel, Vatten (Pakistan). He made important national contributions in the development of computer techniques for the solution of engineering problems – tidal flow computations used in the studies for the Thames Barrier; dynamic relaxation applied to doubly curved arched dams; prestressed concrete pressure vessels for nuclear power stations. In 1966 he was appointed visiting Professor at Imperial College, the first civil engineer to be honoured in this way. He was a leading member of the Channel Tunnel Study Group. He was an outstanding engineer and a man of great charm with wide catholic tastes.

Julian Tritton was President of the International Federation of Consulting Engineers (FIDIC) from 1955 to 1963.

Otter was succeeded by Kenneth Ainscow who was made a Partner in 1969. He was actively involved in expanding the Roads Department until his death in 1976, aged 50. Leonard Hinch, who had been an Associate in the Department succeeded him in 1976 and remained a Partner until his retirement in 1984.

The two premature deaths undoubtedly set back the resurgence of the Roads Department in England which had languished after an earlier difference of opinion with the then Ministry of Transport. RPT had first been appointed in 1938 for a complex underpass scheme, but it was shelved until after the war. Construction started in 1958 and was completed in 1962. Cuerel was the Partner in charge. The contractor succeeded, on appeal up to the House of Lords in a dispute on the measurement of the works. Contemporaries widely believed that as a consequence RPT for a number of years failed to secure opportunities in the English trunk road programme which was then starting to expand. Be that as it may, it is on record that in the late 1950's Strick also turned down the offer to the firm of a motorway project because he considered the Ministry's scale fee was inadequate – by todays standards it would be considered generous! Such are the turns of fate which can profoundly influence immediate future prospects. However, as is recounted elsewhere, the firm's expertise in roads in the UK continued to flourish in work for the Welsh Office, designing and supervising the roads through the difficult terrain of the mining valleys in South Wales.

Newcomers on the scene were Thomas Lambe, who joined the Partnership in 1955, Brian Holloway and Herbert Merrington who joined in 1958 and Peter Fraenkel who joined in 1961.

Thomas Trajan Lambe was essentially a railway mechanical engineer. Initially, he spent some time with the East Indian Railways. He became Controller of Imported Railway Stores, shortly afterwards returning to the East Indian Railways as Deputy Chief Mechanical Engineer and served with the Indian Supply Mission in Washington as a Technical Adviser. He joined RPT in 1947 and succeeded Julian Tritton in 1955. He was acknowledged as an expert on the design of rolling stock, mechanical plant, and in the economics of railway operation. He retired in 1967 and remained a Consultant until, in 1974, he was appointed President of the Engineers' Guild, an organisation established post-war to try to raise the status of professional engineers in the community. Regretfully, engineers are often too devoted to their work to give time to such matters, and the organisation was later disbanded. Thomas Lambe died in 1986.

B.G.R. Holloway – Partner 1958-1975.

B. G. R. Holloway had joined RPT in 1927. He was a structural engineer and expert in steel bridge construction, for many years closely associated with Ernest Bateson, whom he succeeded in 1958. He was involved with both the design and construction of the large steel bridges across the Hooghly at Howrah, across the lower Zambesi and at Baraka. During the War, he served in the Royal Navy, being responsible for special devices used in boom defences. As a Partner he was responsible for projects covering virtually the whole spectrum of civil engineering. He was an extremely popular man and retired in 1975 at the age of 70.

Herbert Merrington was born in Haslemere in Surrey in 1904, and started his working life with Surrey County Council in the Bridges & Highways Department. His subsequent career and work with the firm appears in Chapter 8.

Peter Fraenkel joined RPT in 1945 and six years later was appointed Resident Engineer for the Tyne Improvement Commission's new deep water quay. On return to the office in 1953 he became responsible for many reinforced and pre-stressed concrete structures and general civil engineering works. He joined the Partnership in 1961 and spent some time in Australia setting up the firm's new branch offices. But, he was an ambitious man, and power politics within the Partnership resulted in his resignation when Francis Irwin-Childs became Senior Partner in 1972.

In the six years following the retirement of Cuerel and Lambe in 1967 (the retirement of Merrington and the death of Otter followed in 1969) eight new Partners were appointed. These were the last of the Partners who served an apprenticeship year as Prospective Partners when they attended Partners meetings to learn the roles and were still under observation as to their suitability. Business practice now precludes that more overt process. The first to be appointed was Peter Cox.

In 1968 John Munro who had joined RPT in 1950 was appointed Partner until his retirement in 1985. He served in the Royal Engineers from 1939 to 1947. In the practice he was responsible for many harbour and port development projects including Mina Sulman Port and other projects in Bahrain. He had two long periods away from the office firstly on the site staff for a major port development at Karachi, Pakistan, and, later, for the construction of Fawley Power Station.

David Fairweather – Partner 1972-1980.

John Munro – Partner 1968-85.

David Fairweather, Mechanical Engineer, joined RPT in 1967, and was a Partner from 1969 until he became a Consultant to the practice in 1980. His earlier experience had been with the East African Railways and Harbours, and British Rail (Traction and Rolling Stock). He followed Thomas Lambe in charge of the Mechanical Engineering Department and directed the input of railway technical and operational expertise to the firm's projects. He was responsible for the design of the massive machinery needed for the Thames Barrier.

The Partner in charge of the Mechanical Department also had the oversight of the Electrical Services work of the firm. These services ranged from the lighting of highways, through the powering of pumping installations for dry docks to the operating machinery for the Thames Barrier. Stuart Pratt who had joined RPT in 1953 took over the control of these services when he was appointed Associate in 1971. There were few projects in the firm which did not receive some input from this specialist group. Pratt was appointed a Director of Rendel Mechel in 1986 and retired in 1988.

There were few projects in the firm which did not receiv some input from this specialis group.

The firm's work on Water Resources was focussed again when Edward Haws was appointed Partner in 1978. He had trained with Sir Alexander Gibb & Partners and immediately prior to joining RPT had been Managing Director of Engineering and Resources Consultants. Haws is a member of Panel A. R. under the Reservoirs Act, and sometime Chairman of the British Hydromechanics Research Association and of the British National Committee of the International Commission on Large Dams.

Kenneth Ainscow who joined the Partnership in 1969 and Ralph Downham who joined in 1970 have already been referred to elsewhere.

In 1972 when Francis Irwin-Childs retired, the work load of the practice required that three more partners needed to be appointed.

Alan Fisher, who had spent much of his working life in Hong Kong before he joined RPT, remained until his retirement in 1982. He was involved with a wide range of work.

Brian Luxton who had joined the Ports Department in 1952 became responsible for many of the major port projects. Libya was his main stamping ground and he added Australia to his earlier work areas in Chile and Canada. At home his most important projects were the major marine terminal at Hunterston and the oil platform building dock at Portavadie. He became a Consultant in 1985 when the practice merged with High-Point.

John A.N. Dennis – Partner 1972-1979.

Brian Luxton – Partner 1972-19

The last of the three was John Dennis. He had been with RPT for some 20 years and very much involved with the many Power Stations for which

RPT were the designers. His forte was the design of concrete structures and he was involved with the firm's early development work on offshore structures. He remained a Partner until 1979 and continued thereafter as a Consultant.

In 1974 Peter Clark joined the Partnership, refreshing the Structural Steel and Steel Bridge Department with his wide experience of research and development acquired from his previous appointment as Chief Engineer, Constructional Steelwork Research & Development Organisation, and before that as lecturer at Imperial College. Earlier service with Cleveland Bridge and Engineering Co., on the construction of the Wye Bridge and with Freeman Fox & Partners enabled him to blend theory with good practice. During the development of the gates for the Thames Barrier he had acted as Consultant to RPT. In due course he was to hand over his Department to Directors Richard Tappin (Bridges) and John Dawson (Offshore Structures) and as Assistant Managing Director (Business Development) to expand the firm's work-load into the booming redevelopment of our cities.

Peter J. Clark – Partner 1974-1985 Director 1985-

David Hookway, who was destined to become Managing Director after incorporation of the business in 1985, had joined the firm in 1963. Prior to that he had served with J. R. Smith & Sons, a small construction company specialising in steelwork, Taylor Woodrow Construction and the Central Electricity Generating Board. He was involved on the design of many of the major port projects and was a member of the site team supervising the construction of the Marine Terminal at Westernport, Australia. Prior to his appointment as a Partner in 1978 he made an important contribution the firm's business in Brazil, as already described in a previous chapter. Subsequently he became responsible for railway projects as well, and as a Director of the RPT Economic Studies Group oversaw many projects bringing together all facets of transportation, planning and economics. The Ras Lanuf Port project in Libya again already described, is no doubt the project which he would regard as his most prestigious civil engineering work. His personality enabled him to play a major role in the firm's business development activities worldwide and he was instrumental in the development of the UK Regional Offices through the acquisition of engineering departments of New Town Corporations. As a keen sportsman he shared in the firm's Sports Club activities in squash and cricket at home and overseas. In his earlier years he had added football to his participation in local teams, representing Dorset and Middlesex County.

With the intention of enlarging the Roads and Transportation Department Peter Corfield was brought into the Partnership in 1981. He had acquired wide experience in this discipline with other firms of consulting engineers. He was particularly energetic in business development but left in 1986 and joined another practice.

Throughout the book people named have been predominantly Partners and latterly Directors. Rightly so as it is they who have determined the strategy of the business. However a consulting engineering practice is critically dependent on the relationships its local project managers and resident engineers are able to establish with the clients, contractors, government officials and the people affected by the projects. For significant periods those staff are 'the firm' and their conduct determines its ongoing status. Equally important are the senior staff and project leaders in the office, responsible for work in the UK and overseas who have achieved a rapport with clients and officials which has been vital to the business. Any lists of such names is bound to offend by omitting others who have served the firm and this country equally well. However some post-war senior staff have been named under particular projects for which

Richard Tappin – Director 1985-

Westernport Marine Terminal, Australia – one berth for 100,000 dwt tankers and one berth for 30,000 dwt tankers.

they had a key role and their listing recognises the contribution that all staff have made to the ongoing business of the practice.

Before we leave the personnel of the Partnership era, mention must be made of ladies who, apart from one or two wives of Partners, have been absent from the narrative. As far as can be ascertained the first female employee of RPT was Hilda Streets who joined during the Second World War as a stenographer – prior to that this work had been done by men. She later headed up the letter registry and typing pool which gave the firm exemplary service for many years. Also to note is Jean Post (née Bull) who served as personal secretary to Strick and his three successors as Senior Partner. Now there is no discrimination and female graduates under training serve their due time on construction sites without comment: nevertheless it was not always easy to persuade some of the older resident site engineers to initially accept them, particularly on sites involving night shift work!

Major events described need to be set in the wider context of the changes which were occurring in professional organisations. A partnership is a unique form of business structure, allowing great flexibility in methods of operation. It is best suited to a group of like minded professionals normally working in close proximity to each other. In the modern world it is less well suited to a varied engineering practice spread across the world. By the late 1970's it was evident to some of the partners that changes were needed if the practice was to expand its market share or even survive. Partnerships are taxed at personal tax rates and a crippling top rate of 83% gave no opportunity to build up working capital. Hyperinflation exceeding 25% in one year of that decade made impossible any credible financial planning. Moreover, the equity base was reducing as retiring partners took out their capital which, in the financial climate, could not be replaced fully by incoming new young partners. A

John M. Dawson Director 1985-

service company, Delmerton Technical Services Ltd., was set up but ownership was still with the partners and this limited the benefit which could be derived from that corporate form.

The reconstruction of Victoria Street, Westminster, SW1, which was underway in the early 1970's caused RPT to move from '125' in 1971. Consulting Engineers had traditionally been located in that area as, in the heyday of the development of roads, railways, canals, etc., they had attended on Parliament in the support of their clients to secure the passing of Private Bills which were needed to allow such works to proceed. This is now much less frequent and many firms had taken such opportunity to move to less expensive and congested locations out of the heart of London. RPT decided that they preferred to remain close to the centre of national affairs. After much searching the choice was made – 61 Southwark Street, SE1, on the South Bank. This location provides good access from the London terminal used by most of the firm's rail commuters, and, in spite of its proximity to the City, it has remained largely unaffected by the punitive rent increases which have occurred in more favourable areas. Extensive reconstruction in Southwark is now raising the status of the area but the effect in rent levels has yet to be seen.

David W. Hookway – Partner 1978-1985, Director 1985-1986, Managing Director 1986–

In 1975, a new RPT subsidiary company, the Economic Studies Group, was formed. Originally established in 1969 as a separate department within RPT, it provides economic consulting and financial planning services for the evaluation of public and private investment projects and has built up an enviable reputation for its work in the transportation sector. The Group was initially led by Geoff Ody who resigned in 1985 to start his own practice. Later Robert Smith who had been Deputy Managing Director of the Economist Intelligence Unit was appointed Managing Director.

The staff includes professional transport economists, statiticians, operations research specialists and experts with particular experience of energy generation and transmission, ports, shipping, railways and road transportation, transportation research and land use planning. ESG's capability to provide analyses of the economic environment within which projects are planned and the probable economic consequences of such projects enables commercial factors to be incorporated into project planning and design. To day, RPT Economic Studies Group Ltd., performing as an independent entity within the High-Point Rendel Group, provides a vital service to its principal clients which include the World Bank and other development banks and governments.

As the work of ESG is inevitably at the commencement of a project it became the natural focus for business development activities, the need for which was becoming increasingly important. Soon this function was hived off to become the Business Development Unit which now serves the whole Group. High-class presentations of project brochures, proposals to clients, displays at conferences and exhibitions, etc., are now an essential part of a successful, professional business and the BDU supplies these skills in support of the project departments.

Robert E.G. Smith – Managing Director, ESG, 1985-

Computing has played a major role in underpinning RPT's activities since the brilliant pioneering work of Otter in the 1960s. Dr. Rex Gaisford one of the first specialist Associates, continued these early initiatives by enhancing the reputation of the practice in analytical fields and by establishing the first automated accounting systems. All this early work was carried out at computer bureaux, accessed in later years, by a Remote Batch Terminal. Barry Blunt, now a Senior Assistant Director, succeeded Gaisford in 1975 and has since nurtured the Computer Services Department through a lively childhood to its current adult status. In 1977 he was responsible for installing the first in-house multi-user minicomputer, a

Prime 400. The hardware inventory has now expanded to four Prime and Dec minicomputers along with a population of over sixty microcomputers. What was once a centralised function is now decentralised with Blunt and his small team controlling the core units and major developments, and ensuring that the best practice is maintained throughout the company. It was perhaps appropriate that the Anniversary Year saw the establishment of a Wide Area Network (WAN) which uses modern digital communication facilities, provided by Mercury, to link the hardware resources at the London, Birmingham and Warrington offices. The software library installed on this hardware comprises over 100 programmes applicable to most areas of RPT's activities. Patterns of computer usage change to reflect changing work loads in the firm. For example, the early 1980's saw a boom in the complex computerised analysis and design of offshore structures. Vast structural idealisations, non-linear soil mechanics, and interactive graphics are all used in models which would have been unthinkable two decades earlier. In more recent years, consultancy references related to highways projects have given RPT major exposure to the latest computing technologies such as Relational Databases, Fourth Generation Languages, Expert Systems and Formal Design Methods. In spite of the enormous advances in the last few years Informal Technology is still growing at a fast rate and the coming decades will see the firm using technical, project management and administrative systems which are even now only at the conception stage. Competitive business strategy will harness these technologies to the engineering expertise at the heart of the practice.

The Anniversary Year saw the establishment of a Wide Area Network.

In 1983 RPT invested in its first two Computer Aided Drafting (CAD) workstations. Ronald Morris, who had joined the firm in 1971 and had become a Senior Draftsman, was appointed to develop the service. By 1988 there were five workstations in London, an extensive library of detailed information was available on call and RPT had trained a first class team of CAD operatives to work alongside the conventional drawing office in providing sophisticated drafting services to the firm. In all this the Computer Department was intimately involved.

RPT invested in Computer Aided Drafting.

Not all the Directors welcomed the change and the concentration of much of the drawing production away from their own departments where their engineering teams had direct liaison with the draftsmen. CAD undoubtedly has advantages. Accuracy in dimensions and details is greatly improved, drawings have a uniform style, and changes can be quickly introduced. Disadvantages include the fact that the design engineers are somewhat remote from the 'hands-on' development of their ideas.

RPT, as others, has found that the cost economies expected have not always been achieved. Nevertheless skilled draftsmen may not be easy to obtain in future years and CAD is likely to be the main tool for drawing production. The present equipment is expensive, but reduction in cost of both hardware and software as the information technology revolution goes on will no doubt result in design engineers having direct access to the equipment to oversee their own drawings again. The work done since 1983 in developing these skills within RPT will stand it in good stead to accommodate profitably the next changes.

The work will stand RPT in good stead to accommodate profitably the next changes.

Whilst the major skills of a consulting engineering practice are the planning, design and technical control of the construction of projects its reputation in relation to its client often rests on the manner in which it directs the project contractual arrangements – unless these are correct the projects will not be completed on time and the client will incur unnecessary costs. Unlike an architectural practice it is unusual for a firm of consulting engineers to hand over this responsibility to a firm of quantity surveyors. Project engineers in RPT have always been encouraged to develop contractual expertise so that clients would receive a comprehensive, integrated service. However the legal intricacies of contractual arrangements necessitate specialised knowledge which is provided through the Contracts Department. This small Department, comprising quantity surveyors and engineers who have acquired a particular skill in contractual matters, works in conjunction with the Project Departments, preparing contract documents, advising on contractual matters and, usually, dealing with the claims which contractors, regretfully, expect to make to improve the financial out-turn of their contracts. Donald Cameron headed the Department for many years. He was succeeded by David Masson who, on his retirement in 1986 handed over to Bruce Claxton, Director, an engineer with legal qualification, who joined RPT from High-Point and is referred to later in the Chapter.

A formal association was made in 1978 with Bush & Rennie Associates who were experts in Project Management. Both Gorden Bush and James Rennie (a relation of the famous Scottish engineers of that name) were known to RPT when they were respectively Project Manager and Director with Charles Brand & Co., Contractors, who had carried out the construction of the Belfast East Twin Dry Dock and Carrickfergus Harbour Improvements under the supervision of RPT. Until 1985, when the association was terminated, the company provided to RPT the detailed input of estimating and contractor's operations needed for the economic evaluation of alternative designs and for cost estimates in reports to clients.

A study carried out by a top accounting company in 1979 led to the appointment of a financial director, Andrew Sharp, and the introduction of a new computer based management-finance system to replace the more limited computerised costing system which had operated since the 1960's. In hindsight it is evident that corporate financial advice, even in the realm of sound financial practice, then was much less innovative than it is to day. Consequently the Partnership did not derive the full benefit which could have followed from that study.

About this time RPT formed an association with three leading firms of Consulting Engineers styled MRT Consulting Engineers Ltd. Its object was to promote and carry out work in Latin America and specific African countries. Mott Hay and Anderson were specialists in tunnels, underground railways and large span bridges; Sir M. MacDonald and Partners specialised in irrigation, drainage and water resources (those two companies merged in 1989 to form Mott MacDonald); John Taylor & Sons Ltd (now part of the Acer Group), were the water supply and sewage experts. It was felt that by acting as a consortium they would be better able to secure the mega-projects which were then seen as a likely basis for developments in those areas of the world and to meet the intense competition as developing countries acquired their own expertise. Additionally, national governments were sponsoring consultancy services as an aid to export promotion, and financial agencies such as the World Bank were selecting consultants on a wider basis.

A few small projects were carried out by the individual companies but the mega projects did not develop. Consequently, as a result of some overlaps in expertise between the firms, the association fell into disuse.

The Partnership acquired Transportation Planning Associates.

In 1983, the Partnership, through a contact of Peter Corfield, was able to acquire the Birmingham based firm Transportation Planning Associates from M. M. Dillon, a Canadian firm of consulting engineers. TPA had been founded in 1964 to provide specialised services in the engineering aspects of transporting people and freight. It had developed an excellent track record but the remoteness of the owners had been an inhibition on expansion. The link with RPT was of considerable benefit to both organisations. TPA, continuing under the leadership of Terence Mulroy, now Chairman of TPA and of the Economic Studies Group, had the support and contacts of the RPT organisation: RPT in its turn was able to enlarge the transportation work of its Highways Department to provide a fully comprehensive service. As a result very many more studies, planning and design commissions were received from the Department of Transport and a wide variety of clients at home and overseas. In addition to major projects described in Chapter 12, many smaller schemes such as the Axminster and Salisbury by-pass have been secured. Robin Cathcart is Managing Director of TPA.

Terence Mulroy — Chairman Transportation Planning Associates and the Economic Studies Group.

Overseas a number of corporate entities had been set up over the years. Although control of the operations of the practice was firmly centred in London these overseas entities were run on a light rein. Each tended to be the fiefdom of an individual partner. The entities were seen as a source of project work, rather than as modern style profit centres. Some were established to comply with national regulations often designed to restrict overseas penetration and remittances. Joint ownership with local consultants was the norm. Such companies in Australia, India, Brazil, Iran and smaller ones elsewhere blossomed for a period after the projects for which they had been established were completed but they faded as national restrictions on offshore ownership increased. The recession caused by the oil price explosion in the 1970's was a further inhibition in overseas development.

Into the 1980's further changes were beginning to have a more dominant influence on the practice. Competitive pressures were increasing. As far back as the 1960's much overseas work for the World Bank and other international funding agencies was being secured through competitive bidding rather than negotiation. The preferred system for this was that the client would seek bids from a few consultants selected from prequalification proposals. Bids would be on a two envelope system, one containing the technical proposal, and the other a financial offer. Having selected the best technical bid the client would then open the corresponding financial bid and open negotiations with the selected consultant. This worked well and the reputation of RPT stood it in good stead. However not all clients were sufficiently discerning to appreciate the merit of the system and were selecting purely on price. When competitive bidding spread in the UK public sector, price was the selecting mechanism and starting during the period of the depressed construction market in the late 1970s, consulting firms were submitting cut throat prices to maintain market share. This together with the tendency of clients to take longer credit periods, put severe pressure on cash flow and bank overdraft limits.

Into the 1980's further changes were beginning to have a more dominant influence.

With increasing litigation, individual partners in partnerships became more acutely aware of being sued and, insurance not withstanding, of themselves and their families being stripped of their assets. The corporate sector which provided some protection began to look more attractive. Moreover, bankers and many clients were making it clearer that they preferred to work with a corporate structure than with a partnership. The firm's bankers, National Westminister Bank plc., particularly, appeared to have had little comprehension of the nature of operation of a large international partnership consultancy. The Bank's view of the poor cash flow from the Partnership's extensive Libyan operations was the major factor forcing changes in the organisation and ownership.

RPT had signed its first contracts in Libya, for the ports of Misurata and Derna in 1969 whilst King Idris was still in power. After the revolution which brought Col. Gadaffi to power, RPT continued to work, developing relationships with the new officials. These personal relationships have always been good and inspite of the varying political expressions of the regime, staff working there have suffered no personal animosity. Oil price crises, and other factors, have slowed the remittance of fees making difficult the continuation of the firm's activity. However the Group continues to operate there.

Ian W. Reeves — Chairman, High-Point plc.

Consultancy practice in the developing countries is a high risk business. Changes in political regimes and world trade patterns can delay projects for which staff have been earmarked, and the flow of funds can be impeded as countries seek to control their spending.

All these pressures sharpened the focus on the need to improve the equity base of the practice and to bring in more younger partners. The Partnership Agreement which required unanimous agreement on the appointment of new partners, and on making major changes to it, was proving to be an inhibiting factor in responding to the pressures.

In 1984 Ian Reeves, Chairman of High-Point plc., a person well known to some of the partners, was invited to act as a consultant to advise on corporate and business development strategy. Ian Reeves started his career with RPT. His father had worked many years with the firm and other High-Point people had worked for RPT in the past.

During the period of consultancy, Ian Reeves recalls, it quickly became clear to him that a change of management direction could open up exciting new possibilities for RPT's traditional skills. Corporate people could provide a new direction better suited to strengthening the financial base needed to generate funds to repay retiring partners, finance necessary capital investment, and increase the working capital as required to expand the business.

The Partners therefore initiated discussions with several appropriate parties including High-Point with a view to finding a corporate partner not only compatible but right for the future development and growth of the business. They needed to find a partner which could bring substantial assets to the business and the necessary experience and skills. The conclusion reached was that High-Point was a UK company which could meet the requirements. High-Point might be a smaller company but it had a proven management which had an impressive growth record and expansion strategy.

They needed to find a partner which could bring substantial assets to the business and the necessary experience and skills.

High-Point had been founded in 1970 as a consultancy specialising in providing financial, contractual and management consultancy services to the project development and construction industry. Operations expanded into Europe, the Far East, Africa and, more recently, in the USA. By 1983, High-Point had reached a stage in its development at which the Directors considered it appropriate to float the company on the Unlisted Securities Market and to implement a corporate strategy focused on geographical expansion, expanding the range of services offered and broadening the Group's client base. High-Point Services Group plc. became fully listed on the Stock Exchange in 1987.

... in 1985 the partnership was incorporated as a private company.

The first steps towards a possible High-Point/Rendel merger were taken in April 1985 when the two parties signed an Operating Agreement. Ian Reeves became Chief Executive on secondment from High-Point which also seconded a Financial Director and Company Secretary, Bruce Claxton. At the same time, David Hookway, previously an RPT partner in charge of ports and maritime works, was appointed Director of Operations.

Reorganisation continued and in 1985 the Partnership was incorporated as a private company (Rendels Limited), the shareholders being the seven partners who had previously owned the Partnership. In February 1986, High-Point Design, a wholly owned subsidiary of High-Point,

The Genesis Centre, Birchwood Science Park, Warrington.

entered into an agreement with those shareholders which provided a put and call option for the purchase of their shares exercisable after the audited accounts of Rendels for the three year period to 31st May 1988 were available.

Peter Cox remained as Chairman until his retirement to the role of Consultant in May 1988. However, the day-to-day direction was assumed by David Hookway, who was then appointed Managing Director. He is now also a Director of the main High-Point Group Board having played a key role in the reorganisation of the Rendel Group. In May 1988 Robert Wharton, Deputy Chairman of the High-Point Group, also became Chairman of the Rendel Companies and Ian Reeves relinquished his role as Chief Executive in order to concentrate on longer term business strategy.

Benefits from the link up soon became evident in a number of important aspects. Further younger directors were able to be appointed in RPT and middle managers were given a more transparent status. The grade of Associate was abolished and middle managers were appointed as Senior Assistant Directors or Assistant Directors. New systems focused attention on accountability in operating groups, helping to reduce operating costs and defining risks more clearly. Welding together of the two cultures of High-Point and Rendels developed as they worked together and established important personal relationships at group-wide management conferences. As will be evident later in this chapter the experience of these changes was to be put to good use in aiding the more easy absorption of other entities joining the Group. The closely integrated management structure which continues to develop will provide wider promotion prospects for those with initiative to face the challenges and seek commensurate improvements in salaries and benefits.

Robert V. Wharton — Deputy Chairman and Chief Executive, High-Point plc and Chairman, Rendel Palmer & Tritton.

Proposed atrium development, Galleries Shopping Centre, Washington – an example of the work of RPT's Darlington-based architectural team, formed from the staff of the Washington Development Corporation.

The second main benefit is the ability of the Group to respond quickly to clients who are looking for a total service, blending engineering, commercial and business skills. This is needed not only for projects organised in the conventional manner but for the new generation of design/build/operate/finance contracts being promoted by the private sector.

In parallel with forging the link with RPT, High-Point was expanding in the US market which it entered in 1984. In 1985 it acquired Schaer Associates, a contracts and claims consultancy operating from three

The NEK factory, Washington, designed by the staff of Rendel Palmer & Tritton (North East).

offices on the East Coast of the USA. That business has now been brought together with the Group's original core business, High-Point CTMS, to establish the High-Point Schaer Division of the Group which now operates from 20 offices throughout the world: 5 of those are in the UK and 11 are in the USA. That link will greatly facilitate RPT moving into the US market in due course.

John Forbes Director, Rendel Palmer & Tritton (Wales).

Alun Parfitt — Director and General Manager, Rendel Palmer & Tritton (Wales).

An example of the work of Rendel Palmer & Tritton (Wales) — the Six Bells housing enveloping scheme, Abertillery, Gwent.

The ability of the Group to invest in the reorganisation and expansion of RPT has been of benefit in developing its needed increase in the percentage of its turnover earned in the UK, where in spite of the tight margins already referred to, operations are more stable than in many areas overseas. Historically RPT had permanent offices in the UK only in London. Clients increasingly require consultancy services to be locally based. The privatisation of the engineering departments of New Town Corporations in 1987 and 1988 provided an opportunity to establish regional subsidiaries having an ongoing workload. On this basis subsidiaries were established in the North West and North East following the acquisition of those departments at respectively, Warrington and Runcorn, and, together, Washington, Aycliffe and Peterlee. Thomas Walsh who is Managing Director of RPT (Warrington) is also an Assistant Managing Director of RPT. In South Wales the department at Cwmbran was consolidated with Parfitt and Forbes, a consultancy practice established by Alun Parfitt and John Forbes which had just joined the Group. Both founders remain as Directors, Parfitt being the General Manager, RPT (Wales). Each of those entities brought new markets within reach of the

The Warrington North Effluent Treatments Works designed by the engineers of Rendel Palmer & Tritton (Warrington).

Rutherford House, Warrington — a reinforced concrete framed office building by the structural engineers of Rendel Palmer & Tritton (Warrington).

Thomas Walsh — Managing Director, Rendel Palmer & Tritton (Warrington).

Group and, importantly, added expertise in building and private industrial development which had not featured strongly in RPT's heavy civil engineering practice. In Birmingham a subsidiary company was established to work with Transportation Planning Associates which, from its home base there, had developed major work on road schemes. John Smith, Director, leads the RPT (Midlands) team.

Expansion also took place in the range of Rendel's services. The wider Group has enabled Transportation Planning Associates, and the Economic Studies Group, to be substantially enlarged in revenue terms. In-house departments providing geotechnical services and electrical/mechanical services, which are essential requirements for most major projects have been formed into subsidiary companies able to offer services in their own right. Rendel Geotechnics, headed by Dr. Alan Clark, Managing Director, has been strengthened by the acquisition of Geomorphological Services Ltd, whose Managing Director is John Doornkamp. The electrical/mechanical department of RPT is now part of Rendel Mechel, the latter name having come from a High-Point subsidiary already operating in that sector. Michael Stout recently succeeded Alistair Duncan as Managing Director. In its turn the latter company has been strengthened by the acquisition of Hancox and Partners, A Glasgow-based consulting mechanical and electrical practice. This link has also given RPT a regional base in Scotland. David Carmichael is Managing Director of Rendel Hancox.

The largest and most recent acquisition to date has been that of London Scientific Services. This little known organisation was established over a century ago to provide scientific services to the London Boroughs and other public sector organisations serving the capital city. It is headed by Samuel Radcliffe, Managing Director. It employs over 100 staff and operates laboratory facilities equipped with modern sophisticated equipment. Its acquisition following privatisation will enable Rendel Scientific Services to expand worldwide a wide range of analytical and consultancy services for environmental impact analysis, pollution monitoring, and the assessment of quality of building materials, components and other goods. Those services will be complementary to those offered by Rendel & Branch, landscape architects and designers, a practice which had already joined the Group. The last acquisition neatly

Water quality monitoring by the Group's consulting scientists, LSS.

Sam Radcliffe – Managing Director, Rendel Scientific Services.

Simon Rendel – Director, Rendel & Branch.

rounds off this chapter, but undoubtedly not the expansion of the Group, because the Rendel in the name is that of Simon Rendel a great-grandson of JMR, the founder.

12
Into the 21st Century

Technological innovation is the life force of a consulting engineering practice.

Technological innovation in anticipating and meeting the aspirations of clients is the life force of a consulting engineering practice. The history of the many projects of the firm set down here shows that it has not been found wanting. In the post-war period commercial and bureaucratic aspects have also emerged more strongly as important factors in the life of the business. In looking to the future in this final chapter current projects illustrate the complexity of the situation in which the practice now operates.

The strategy of deregulation and more often open competition adopted by the UK Government and others, has led to wider demands for full service organisations able to offer worldwide financial and commercial services together with a wide range of technological expertise. The latter includes with main stream engineering the closely related disciplines of transportation and environmental planning, economics and social matters. Clients recognise that such services are best provided through one lead organisation which is able to call on particular specialised disciplines with confidence in their experience and their ability to respond positively to people they already know. People are the essential assets of a knowledge based profession even though information technology is enabling computers to take over much of the more routine work. This multi-service function is offered by the companies which, pre-eminent in their own right, now form the High-Point Group.

The end of the three years after the incorporation of Rendel Palmer & Tritton saw the departure or retirement, from an executive position, of five of the seven Partners who had been the last owners of the business; David Hookway, Managing Director, and Peter Clark, Assistant M. D remained. Edward Haws, Director, continued as head of the Water Resources Department. Promotions to Director of long-serving personnel and the recruitment of one Director were made to maintain continuity and meet the business challenge.

David V. Hattrel – Director 1985-8, Assistant Managing Director 1988-.

The biggest changes have been in the Roads Department which has expanded significantly in the last few years. David Hattrell who had joined RPT in 1965 was appointed Director in 1985 and Assistant Managing Director in 1988.

Management and maintenance of highways to cope with the burgeoning growth of traffic is of major concern to its clients. David Hattrell has led RPT in the development of comprehensive computer-based management information systems. Work was done in collaboration with Hertfordshire County Council following the vision of their late County Surveyor, Michael Hardy OBE. He perceived the need for an overall management package ultimately intended to encompass routine and pavement maintenance, traffic, accident, public utilities, bridge inspections and all facets of the duties of a Highway Authority. Under David Hattrell, RPT provided the engineering and systems know-how to turn this vision into a reality. The HERMIS system is now operating in Hertfordshire and marketed internationally jointly by the County Council and RPT. Recognising this particular expertise, the HERMIS team were also invited to undertake a commission from the Department of Transport and all local authorities in England and Wales to design a road pavement management system for use on all of the country's roads through into the 21st century. It was a tribute to the team that the forward looking, imaginative and innovative design finally produced, was backed up with a pioneering cost-benefit analysis which demonstrated quantitatively the worth of introducing the system.

The logo used in the marketing of RPT's Highway Engineering and Road Management Information System (HERMIS).

Equally challenging has been the study for the Department of Transport of the M25, London's 180 km long orbital motorway, for which Michael Kelly, Senior Assistant Director, has been responsible. Increasing congestion requires short, medium and long term ideas for its relief. Absolute solutions are not possible and in preparing the report RPT have to balance the conflicting interests of the users, the environment and acceptable costs.

James Hyde became a Director in the Roads Department in 1985 shortly after he had joined the firm. His earlier experience had included six years as Engineering Adviser to the Overseas Development Administration in S. E. Asia. In addition to the A 449 Western Orbital Scheme for the Birmingham conurbation, which has already been described, his work overseas illustrates three very different types of project.

In Ghana, where RPT had carried out a lot of work in the period before the revolution, a design study is underway for 54 km of road between Nobekaw and Bediakuknom. Finance comes through a competitive bid to the European Development Fund, a fund set up by the European Economic Community to provide development resources to the ex-colonial territories in Africa, the Caribbean and the Pacific.

James R. Hyde — Director 1985-.

In North Yemen, the World Bank is financing the upgrading design of the Sana'a to Hodeida Road. Again this was secured by competitive bidding. Spectacular escarpments have to be crossed as the road undulates from sea level to 11,000 ft above.

Upgrading of the Sana'a–Hodeida Road in the Yemen Arab Republic.

Construction of the Gumusova–Gerede motorway in Turkey under the supervision of RPT.

Turkey sees RPT involved in a complex road project secured after one unsuccessful bid during which lessons were learned about the intricacies of Turkish financial and bureaucratic processes. Persistence in penetrating a new market can pay off. The firm in association with Yuksel Proje, local consultants, was shortlisted to bid for five separate sections of the Anatolian Motorway, the arrangement being that if a firm secured a section it then dropped out of the remaining bids. RPT secured their first bid for the Gumusova to Gerede section of the road – in which they are required to check the design by a contractor (who has also arranged the finance), and supervise the construction. A delicate balance of interests has to be managed alongside the technical requirements. The road is 121 km long and transverses a flood plain before entering a seismically active mountainous region. Four tunnels total 9 km in length and there are extensive viaducts.

The Ports Department is now headed by Anthony Brinson, Director, who had joined RPT after worldwide experience with another firm of consultants. He is supported by Senior Assistant Directors John Warmington and Brian Reeves. The latter also has responsibility as training officer for the firm's graduates under agreement for their professional training.

Anthony G. Brinson – Director 1985-.

Since the early 1970's RPT has had an association with PT Indulexco, a firm of financial and engineering consultants, in Indonesia. In 1977 RPT, the Economic Studies Group and the local associate were engaged to study the economic and engineering development of the Port of Surabaya

Surabaya Port, Indonesia. The photograph shows the construction of Phase Two of the development which includes two offshore container berths linked to the mainland by a 1 km long piled access bridge.

Artist's impression of the proposed Fawley 'B' power station superimposed on a photograph of the existing Fawley 'A' station.

under finance provided from the Asian Development Bank. In due course the study led to the design and supervision of construction of new general cargo berths and facilities. As with a number of other projects the design and drawings were prepared in Surabaya by a local team under RPT direction.

Subsequently, as the need for container handling facilities developed, a further study was made. Soft ground conditions and siltation caused RPT with the support of Rendel Geotechnics, to propose two berths for container ships in the deep water channel connected to the container storage area by a piled roadway 1 km long. This arrangement, which goes against convention in layout, caused much controversy with the client and the ADB. However, it was eventually accepted as the only sound technical solution and is now being built. David Coode continues as RPT representative in Indonesia, a position he has had for some years.

At home a very different sophisticated technology has gone into the design of the floating facility at Coulport for the loading of the Trident submarines. A fixed structure was shown to be uneconomic on the steeply sloping hard rock loch floor. A 'U' shaped pontoon 200 m long by 80 m and displacing some 70,000 tons into which the boat travels, supports a 15 storey building carrying heavy duty travelling cranes. The client, the Property Services Agency, required tenders to be invited for both a steel and a concrete pontoon and so designs were prepared for both. The tender for a concrete pontoon was accepted as being the most economic, and RPT is now supervising the construction of the whole facility.

Meeting the increasing demands for more energy has been a mainstream activity for RPT.

Meeting the increasing demands for more energy has, with transportation, been a mainstream activity of RPT. Recent demand and organisational changes in the UK have changed the nature of client demands. South Denes Power Station was designed by the firm in the 1960's as a coal-fired station. RPT is now in the planning stage of converting it

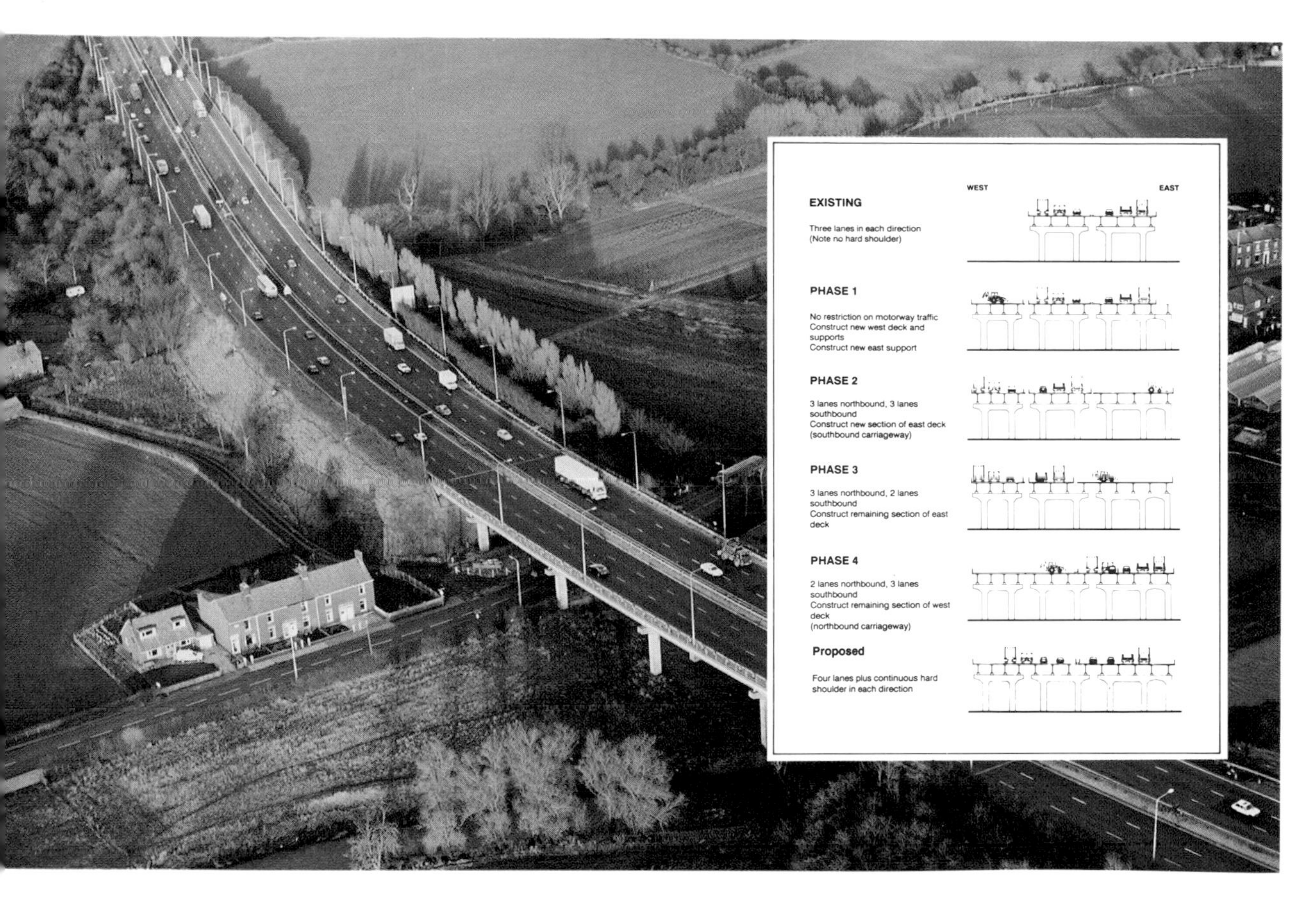

RPT was appointed in 1987 to design a widening scheme for the M6 Motorway between Junctions 30 and 32 near Preston. The photograph shows Higher Walton Viaduct — one of twenty to be widened or rebuilt. The minimisation of traffic disruption during construction (see insert) is critical in ensuring the viability of the scheme.

to a more efficient gas turbine combined cycle station generating 340 MW. The new client is Ranger Oil seeking a profitable use for its North Sea gas. London Scientific Services is working closely with RPT on the environmental impact with particular reference to acoustic assessment and monitoring.

The Central Electricity Generating Board has plans for the construction of a new generation of nuclear power stations. Because of the size and complexity of such stations the CEGB considered that consortia of consulting engineering firms were needed to meet their programme. RPT formed such an organisation with Merz & McLellan, and James Williamson & Partners, firms with whom they already had links, confidently expecting to be responsible for such a station. Reaction to problems of nuclear power and its cost for power generation forced a rethink of the programme but the consortium was engaged to plan and design a conventional 1800 MW coal-fired station, Fawley 'B', alongside the first station previously engineered by RPT. Work was well underway when the Government's intention to privatise power generation in two companies slowed that programme also. However the consortium has secured the generic design of the boiler and turbine houses which can be used on any such station in the future. Consulting engineers regularly respond to rapid changes in clients' thinking but heavy demands are placed on the organisation and on personnel who may have to move house to meet the requirements.

Two obsolete power stations, now sites for redevelopment, also form part of RPT current workload – Shoreham and Brunswick Wharf, on the River Thames. Both these stations and South Denes require the careful removal of asbestos now recognised as a general health hazard and RPT has developed a highly specialised team able to engineer and monitor this work. A major development will follow at Brunswick Wharf.

Urban regeneration is an increasing work load to which the wide

The proposed Exeter Skypark — a 200 acre retail and business park and leisure centre.

diversity of skills in the Group is well able to respond. Two examples will suffice to illustrate this work which is directed by Peter Clark. At Springfields, Newcastle under Lyme, an old clay pit filled with assorted material is being developed for a private developer to provide 300,000 sq ft of retail and leisure facilities in pleasant surroundings. Rendel Geotechnics is dealing with the complex foundation problems and Transportation Planning Associates, the transportation requirements.

At Exeter Skypark another planned development on 200 acres aims to provide a retail and business park and a leisure centre with water sports. Direct links to the M4 and British Rail will be provided. Work started on this project some four years ago with major input from TPA and the Roads Department. The results of the planning enquiry, completed 18 months ago, are eagerly awaited.

E. T. Haws – Partner 1978-85. Director 1985-

In Thailand, Edward Haws is extending the RPT experience of the Thames Barrier to work with an Austrian consortium to put in place for the Bangkok Metropolitan Administration a flood control scheme on the Chao Phraya River which will eliminate the disastrous floods which regularly plague Bangkok, and protect 1400 sq km. A control structure will divert the rainfall run off into a new Diversion Channel to discharge below the city at the Sea-Barrier Structure proposed to protect the city against tidal surges. Locks for shipping and pumps to maintain the river discharge at high tide will complete this complex project. Soft ground conditions demand a major input from Rendel Geotechnics to solve the foundation problems. The Austrian and UK Governments are keenly interested in supporting the project. RPT, working closely through the British Embassy, is looking to financial backing for the project through the Project and Export Policy Division of the Department of Trade and Industry. Nicholas Reilly, Senior Assistant Director, and Alan

The Rion-Antirion Bridge across the Gulf of Corinth, Greece.

Mitchell, Assistant Director, have responsibility for the projects in the Water Resources Department.

In the Offshore Department John Dawson is developing the role of the Offshore Certification Bureau of which RPT is a joint owner. OCB is appointed by the UK Secretary of State for Energy to issue Certificates of Fitness for offshore installations on the UK Continental Shelf. OCB is the only such organisation based on consulting engineering companies, and of UK based companies only Lloyds Register of Shipping has similar authorisation. This venture will be the focus of offshore engineering activity for RPT in the immediate future in a market showing signs of major recovery after the difficult period which followed the oil price collapse in 1985. OCB offers certification, verification, reliability, safety and quality assurance services, and is extending the supply of such services elsewhere at sea and on land where private clients and governments require independent certification of structures and processes.

OCB is appointed by the UK Secretary of State for Energy to issue Certificates of Fitness.

Bridges continue to be important aspects of the firm's activity. Richard Tappin, Director, has the responsibility for this Department. Senior Assistant Director Graeme Marshall, directs particular projects.

Replacing a moveable bridge at Stoneferry, Hull, for the Humberside County Council is the latest of many similar projects currently in hand.

Logo of the Offshore Certification Bureau.

In Greece, in association with Eupalinos Technical SA, RPT is assisting the Ministry of Public Works in evaluating bids for the Rion-Antirion Bridge. The crossing is 2.3 km over a waterway 70 m deep used by many large vessels. The RPT design provided a suspension bridge with a 1450 m main span formed of an aerodynamically shaped steel box girder. Pier movements of up to 2 m under earthquake conditions have been allowed for. It was some 18 months after RPT had put in a bid to prepare a study for the bridge that they were asked to prepare quickly the outline design and bid documents. This they did in two months. The bid evaluation has been underway for 12 months.

Perhaps of all the projects currently underway the one which has brought into action so many of the Group's capabilities is the Jamuna Bridge in Bangladesh. That country of 110 million population with a GDP per head of $150/annum is divided by the great Jamuna River. Two vehicle and one train ferry cross it but the shifting river makes these slow

Artist's impression of the proposed 5 km long Jamuna Multi-purpose Bridge in Bangladesh which will create the first fixed transport link across one of the world's great rivers.

and less than sure links. The western half is further divided north and south by the Ganges River which joins the Jamuna and is crossed only by the Hardinge Bridge constructed to RPT's design in 1915.

In early 1986, after RPT had successfully completed the Power Interconnector already described, the firm, in association with Nedeco, Dutch consultants, and Bangladesh Consultants Ltd., secured a bid to the World Bank for a study funded by the United Nations Development Programme. The Phase I report comprised a characteristics and configuration study which concentrated on the location of the bridge and on a broad range of technical and economic issues regarding the multipurpose nature of the project. This was followed in Phase II by a feasibility study which addressed in detail road and rail links, electricity and gas interconnectors and more broadly the issue of Bangladesh railways and the consequences on the rail network (broad and metre gauge) being included or excluded from the bridge river crossing. The reports have been the subject of many discussions with the Jamuna Multipurpose Bridge Authority, the World Bank, other international agencies and a team of six international experts appointed to review the studies. A decision on the scheme is due early in 1989.

Transportation Planning Associates have carried out the traffic forecasts and the economic studies have been led by the Economic Studies Group.

What is established is that, with a bridge in place, by the year 2020 passenger movement will have increased ten fold; freight will have grown four fold and the transport cost will be less. The country has enormous gas reserves in the east, and agriculture products grown in the north-west provide export potential as well as being able to support the country in food. Efficient transport is a key to raising the desperately low standard of living. Hard economies have to be followed in justifying an expenditure of some $500M on a project of this type but forecasting the ultimate overall benefits of a fixed link demands vision as well as technology.

. . . with a bridge in place by the year 2020 passenger movement will have increase[d] ten fold.

The culmination of a year of celebration: Rendel Palmer & Tritton's 150th Anniversary Banquet at the Guildhall, London. The guest of honour was the Rt. Hon. Paul Channon MP, Secretary of State for Transport.

The engineering problems have been no less daunting as the river is 14 km wide at the site chosen, scours some 45 m deep and changes its course with the monseen flood and the snow melt from the Himalayas. The river will be trained starting 9 km upstream of the site using the town of Sirajganj on the west bank and protection on the east side as hard points. Groins, and massive sand and rock training banks 2.5 m long, will constrain the river to go through the 4.8 km long bridge of multiple 100 m spans. Earthquakes have to be allowed for. With Rendel Geotechnics' advice RPT has established that steel piles will have to penetrate 90 m below the flood plain. RPT schemes presented to the tenderers propose steel girders with concrete decks of up to 2800 tons to be lifted into place; alternatively prestressed concrete segmental units can be used. Any of these units could be made locally or brought by sea from distant fabrication yards. The contract documents being prepared by RPT, with assistance from High-Point Schaer, will invite contractors to submit schemes suited to the world wide resources which can be brought to bear on such a project. In due course RPT and its associates expect to evaluate the tenders and project manage the construction.

* * *

The projects outlined in this chapter illustrate the diverse range of clients and relationships which are now central to the activities of a major consulting organisation such as the High-Point Group. The need to have readily available the skills of all the companies in the group is evident. Increasingly major consultancies have the ability to identify opportunities, to conceive projects and to bring together the organisational strengths – financial, human and technological – needed to act as project promoters. They are better placed to develop their organisations to undertake these project activities, than are contractors and construction companies, both of which have different problems in maintaining this rôle alongside their essentially practical skills.

The nations of Europe, North America, Japan, and others rapidly joining them, are generating wealth at an unprecedented rate. Living standards are rising and increasingly there is a flow of money which can go into private consumption – housing, more leisure pursuits, travel, sports and extra health care – and into the general infrastructure which is the physical basis of civilisation. Increasingly the community is demanding higher standards for the environment and there is no reason why this demand cannot be met. Much of the infrastructure will remain in the public domain and engineering services will be procured in the way they have been in recent years. But with the increasing desire of governments to see privately financed infrastructure and regeneration of our cities, there is a need and opportunity for professional project promotion and development, and for a return to the traditional rôle practised by the great entrepreneural architects and engineers of the 19th century – amongst whom are numbered JMR and his early successors – a period during which much of today's transportation, drainage and service infrastructure was engineered and constructed by private enterprise.

The 150th Anniversary of Rendel Palmer & Tritton has been celebrated with many functions for staff and guests in the United Kingdom and at main centres of the Group overseas. These occasions have served to focus the coming together of the companies which, since the incorporation of the practice, comprise the High-Point Group. The synergy is now evident and James Meadows Rendel, a master of the technology available when he set up his practice in Westminster in 1838, would be proud of his successors as they work with a vast storehouse of expertise which is theirs to command. The next 150 years will undoubtedly see changes equally as dramatic, but meanwhile the men and women in Rendel Palmer & Tritton will continue to offer their services to the worldwide community as they have done in the past.

James Meadows Rendel would be proud of his successors as they work with a vast storehouse of expertise . .

* * *

Index

Note: page numbers in italic refer to illustration captions.